Holy Grail and Modern Science as Duality

Holy Grail and Modern Science as Duality

Olimpia Nera

iUniverse, Inc.
New York Lincoln Shanghai

Holy Grail and Modern Science as Duality

iUniverse books may be ordered through booksellers or by contacting:

iUniverse
2021 Pine Lake Road, Suite 100
Lincoln, NE 68512
www.iuniverse.com
1-800-Authors (1-800-288-4677)

Because of the dynamic nature of the Internet, any Web addresses or links contained in this book may have changed since publication and may no longer be valid.

The views expressed in this work are solely those of the author and do not necessarily reflect the views of the publisher, and the publisher hereby disclaims any responsibility for them.

ISBN: 978-0-595-47503-2 (pbk)
ISBN: 978-0-595-91774-7 (ebk)

Printed in the United States of America

Holy Grail and Modern Science should never be under one roof?

In other words, religion and science can't coexist in harmony as a duality?

Should our world find their harmonic co-existence? Should one play a lower role than the other?

Eventually religion tried to exclude science and science tries to exclude religion.

This superiority role-playing doesn't seem to be an apparent benefit to their harmonic co-existence or to the wellbeing of any social system.

How can one see these two aspects as a unifying force in any civilization?

Never happen before. Will it ever happen?

Their separation will always lead to war and hate among people.

This book is not of philosophical nature, or trying to introduce ideas that did not exist before.

Everybody knows the ideas that I intend to present, but very rarely one would see the link that brings together the past and the future the way it will be presented in this book.

Holy Grail by its recent definition in Da Vinci's code was considered to be rather the bloodline of Jesus. An older version used to this day by Christian church is—the cup—Jesus drunk his last vine on this earth as a "mortal". If one reads more about what Holy Grail could mean, might find endless interpretations, but none will satisfy an inquisitive mind.

Opinions of more fictional nature claim that Holy Grail is: *"the three Grail threads are 1) the DNA, 2) restoration of nature through the Elohim, and 3) the final restoration of the higher strands of the human genome, which are currently in disuse"*. This is according to; "just Wesely". He is someone very much interested in alchemy and "ancient science" and loves to chat about it. And there are many other opinions of this nature that were expressed over time about the Holy Grail and its meaning.

Let's say all possible interpretations are not sufficient to explain what Holy Grail really means;
Should one give up?
Should one persevere?
Should one search to exhaustion?

What if, "by mistake", one just finds out what Holy Grail really is?
What if doesn't want to believe it as true, since, everything what was said or thought before is completely the opposite of what was just found?

A brief explanation to why the Jesus' bloodline made such a big noise for less than 24 hours would be that nobody really cares. Everything is entertainment or moneymaking "faith". People were able to make a market place from the house of God back in Jesus times as they do in our times.

If indeed Jesus had children with Mary, should they be next in line for royal treat? That's exactly the reason why Judas betrayed Jesus. He wasn't the royal he wanted him to be. If Jesus wasn't the royal that he "was supposed to be", how could his bloodline be a royal line?

If anything, if Jesus and Mary Magdalene really had children, they should at best be spiritual leaders of some sort. Like a famous healer, or water walking human, not the Caesar, Jesus refused to be.

Having children doesn't make a human worthless for the love of God. So if indeed Jesus had children that should be a sign he loved God and His ways towards human nature. But in his case as I'll explain later that was an impossible status.

Enoch, Moses, David, Abraham and many other prophets who were God's favorite people were married, had families and still lived in spiritual ways. Jesus was indeed an exception.

Holy Grail as being the bloodline of Jesus doesn't make a lot of sense as it was presented in the "Da-Vinci code". The bloodline of Jesus, by now at it's about 80-100th generation, (if we take an average per generation of 20-30 years) is far from being "pure". Even if his bloodline was pure, his descendants should be the most famous spiritual leaders. Lama seems to gain more leadership in that regard not the ones in some particular French descent as mentioned in "The Da Vinci code".

Jesus wasn't the kind of king Jews were looking for. He was the spiritual king to humanity. A spiritual king will never desire the power of a "regular" king among humans. Such power can only lead to estranged ways from God. Judas was hoping to see the same kind of king as David was. King David made too many mistakes to be a spiritual king.

In fact, if one recalls, the days Jesus had to spend in temptation with Satan, he offered him to be the king of the world, but Jesus said there is more to life than such power. Jesus rejected the power all Jews wanted for him. The power that opens the door to riches and fame was the same power every human dreams of, but Jesus wasn't looking for such power.

What would be the point that Jesus' bloodline would be the kind of kings he rejected "the special offer" to be?

If indeed Jesus had children with Mary, what would they really be?

The kings Jesus rejected the offer for, or the seekers of truth and love he couldn't stop talking about?

Plus, as I recall, God didn't even want for Israelites to have the kind of king other nations had. That's why Israelites started with judges and prophets only.

The idea of small state is not new. God did suggest that idea to Israelites long before our "modern" idea came along.

Since all the nations around them had kings, at some point they did ask for a king. Saul was their best choice. When Saul did forget about spiritual life after he got the power, David was next in line. Over time, even David bought into soul poisoning idea of power and fame and tried to embellish himself in the false idea that all belongs to him.

If Jesus is from the line of David I could see how his bloodline might go back to what David failed to deliver.

Indeed being a *spiritual king* is the best for both worlds. Throughout history never was such king. In fact God did manage to keep these two social aspects separate.

A king of human world was never his own spiritual messenger or prophet.

Kings who believed in God did rely on prophets in their actions. Not even king David was his own prophet.

So if Jesus was a prophet couldn't be the type of King, Judas or the Israelites wanted him to be.

Almost makes sense to why Jesus did not value the offer to be a king of this world.

Having religion and state separate, still is the best solution. God forbid of a king who chooses to control spiritual life for his own power.

Many politicians try to corrupt the spiritual life for their gain. By using the thirst of "feeling important" people have, religious leaders gain the most wanted power. Such power is contagious and can easily be used to manipulate the masses by controlling the political arena or the economic aspects of social wellbeing.

Power and riches have the ability to destroy our spiritual nature by allowing corruption of the soul. Jesus knew that side effect. He rejected that as a very destructive force. So let's see what the world made of King Jesus, who actually was offered by Satan the power to control the world and its riches or be the King of the world, but he rejected that offer.

One of the most intelligent philosophers ever lived on this planet was Paul and he came to the conclusion that one is better off if talks about Jesus resurrection that try to "kill" it. His letters have more knowledge/science than any other philosopher. No other mind that lived before or after him had that kind of scientific information. It is still a great mystery of how comes he knew so many scientific principles that were discovered only in the 19th century by our modern science. At the same time if one analyzes Paul's letters closely, there still is a lot left for our science to "discover".

"By mistake", (I don't believe in mistakes anymore) I found out why Paul tried to get rid of Jesus' followers. When Paul realized that Jesus' teachings are far more influential than he estimated, he used the "long term" strategy. Long term strategy in this case would be;—forget this generation, they will die anyway, rather make sure that the next generations *never get it right.* Even though Paul lived during Jesus' times, and tried to deal with Peter and his followers, he writes his letters in a rather short period of time after Jesus. In fact, so short, that the people he talks or writes to them, still remember him as Saul, as the one who killed Jesus' followers few years back.

At the same time his miracle is not really a miracle. Being blinded by light is not what I would call a miracle. Seems no witnesses wrote about it in the first place, beside him. Why did he have a "negative" miracle, instead of being cured of his dark ideas of destruction? The light he saw blinded him and a healer could heal him. That sounds like a spooky miracle, if you ask me. Of course I am talking in the fashion that Jesus did perform miracles, with lots of witnesses and always happen only by request of others. Also never was a negative miracle, like blinding someone, of course if we exclude the first one. To change water to vine, that did blind many at once.

But let's try to stay with Paul for a bit longer, since he is the one who knew about the Holy Grail and its science.

Paul or Saul made it to "Great philosophers" among Plato, Aristotle and another few minds of the kind.

Paul's legacy was to change his bad image from being the anti-Christ to being Christ's most devoted apostle.

He talks about Jesus' life and his spiritual life in terms no other human on this planet ever did.

That's the reason why not in his time or in our times, Paul was or will be, fully understood or approved of.

Why was Paul so eager to change his attitude after he saw "Jesus" in the light? Did he not know who Jesus was while he was alive? He knew about Jesus' resurrection when he started to persecute Jesus' followers, maybe he did not shake hands with Jesus, but for sure he knew about his behavior and his resurrection.

In his letters is more than obvious that his knowledge wasn't achieved over night. Jesus never spoke about some of the knowledge he did present in his letters. Sure, there is a reason to that, and pro and cons could be dig up to infinite.

As one reads Paul's letters will realize that he had a source of knowledge greater than anybody before him or after him. The question is; why did he try to hide it? Out of fear from the church? Could it be the jealousy, of not letting others know about it?

When he saw that Jesus revealed that information to so many "outsiders", Paul tried to stop them. Peter was his first attack, since Peter, even though, knew a lot less than Paul, was very determined to pay for his sin of betraying Jesus when he was killed by Romano-Judaic complot. I am talking about the leaders, not the masses who were "controlled" by religious or political powers.

Paul knew that if he can stop Peter, all the others were too weak in their faith to continue. Peter, on the other hand knew what he knew, even if he couldn't explain it he believed it. That was Paul's tough rock. He couldn't destroy Peter and his followers. At Antioch for the first time, Paul used the term "Christians" when referred to Peter and his followers. That name was their stigmata and lead in the end to their death as *Christians*. Therefore the Jews were in fact the first true Christians, but thanks to Paul they were mislead too.

Paul did realize that the movement is too strong already, and nothing could stop them only if they have some misleading information over time.

So before the true Christians had the chance to gain "terrain" among gentiles (genetes= nations), the nations had to be mislead so they will never fully understand who Jesus really was.

The most attractive option to mislead was to give the impression that Jesus death saved everybody, therefore sin doesn't exist anymore.

Many explain this by commuting the sacrifice of animals to God for various sins. It's true, God did ask of Israelites to sacrifice animals in exchange for their sins. Nobody can really explain why that was. A simple explanation is the human nature itself. We as humans need to pay a price to learn a lesson. For someone who produced an animal of good quality, to have to sacrifice it to God was a great price to pay.

Usually when we pay for our mistakes we learn not to keep making them.

Most people after they pay a fine for making a mistake usually will try, if possible, to avoid making the same mistake. Of course I wasn't talking about parking tickets in NYC.

By killing Jesus, or taking him away from people, that was a punishment only for few, not for most. That was a punishment in the sense that people won't have the great healer and spiritual teacher among them anymore.

Jesus' death did not absolve humanity of its mistakes/sins. That is what Paul was trying to sell to some extent, so people will rely on that salvation, and do not care to learn more about Jesus and his teachings.

Jesus death, to Paul, was more like saying; since Jesus died for you, you are saved and your sin was taken away, so please do continue to feel free to sin.

That was very attractive to most. In fact to this day the church still sells this idea.

Is that true?

If Jesus died for our sins we can continue to sin with no remorse? If is true, why did the church come up with confessions and indulgences?

By now, 2000 years after Jesus one can clearly say that every religion got corrupted in some way, and did not help people to gain true faith. More often than not the church leads people to a lot of distress and shame. Many went insane over religious ideas. To this day our civilization is still under that curse.

I would like to call it "Paul's curse" for now. Vatican embraced Paul's curse.

Of course Paul knew a lot of things which Peter and his followers had no clue about. In his letters he gives away small parts of that knowledge.

His letters have a grain of truth covered in a lot of misleading layers.

In fact by now I think his letters are a wicked gift to humanity.

In a figurative way, he gave away a very important seed, after he boiled the seed and wrapped it in the most luxurious box.

While giving it away he said; "take care of this, is the most precious thing a lover of Jesus can have".

To understand why Paul did that, one needs to have a lot of knowledge and the right knowledge. He knew that knowledge can't be found in one book or in one place, but in a grid like network. Jesus also knew about this grid, but he tried not to mention about it.

That's why Jesus did not write, but delivered the message in deeds not words.

More like a modern scientist would produce the reaction in your face. We would see its effect, but never really understand how it works.

Paul on the other hand, did not know how to produce the reaction, but he knew only a lot of theory about it, but wasn't able to deliver it.

He knew the theory before Jesus came to be Christ. Paul was rather jealous that Jesus found all the dots of the grid before he did. Or maybe he had the entire grid, but didn't want to share it.

Which one is true only the future will tell.

What kind of grid am I talking about?

Recall Da Vinci's code?

The information wasn't supposed to be found in one pot of gold, but rather in a grid like system.

If one had only one part of it couldn't really connect the dots so had to find out the missing parts of the puzzle. Of course the way it was presented in that book, was just a cheap joke.

Our modern fiction and our fairy tails are filled with the idea of grids/networks.

In fact we still joke to this day about it, in our way of wrapping gifts. (Big boxes small gift.)

When I read Paul's letters I was amazed to see that he knew more modern science than we know right now in 2007.

Paul knew the law that Lavoisier discovered back in the early 1800s or so. Antoine Lavoisier was a chemist, and for the first time he did express the law;—"that in a chemical equation/reaction: nothing is lost and nothing is gained but everything changes/transforms".

As soon as he came up with that idea, a lot of things in chemistry started to make sense.

This law of transformation proved to be correct in many instances. Is still one of the most potent laws, but is not applied to all fields yet.

Paul in one of his letters claims exactly this transformation, only the scale is way bigger than our modern science can envision it at this time.

The only problem is that Paul covered that idea so well, that nor his generation nor ours can fully get it yet.

Holy Grail or the "whole—integral grid" was something Paul knew about.

He claims he needs to travel allover the place to all the nations. He changes his plans suddenly, with almost no reason at all. If one follows his trips on the map and what he says about them and the reason why he made the trip, one starts to have doubts about him or just simply think he is just a crazy person.

In fact in few of his letters he claims that maybe people see him as crazy, but in fact he just wants to share what was given to him.

So what was given to him in fact? Did he travel just to create the grid, out of the Holy Secrets, which he claims he has?

Without doubt Judaic church had a lot of information on their hands, a lot of Holy secrets, and if all those would fall in the hands of

Romans or Corinthians or Ephesians or even Galatians would be a total disaster for life on earth.

Imagine, if in those times, Rome would have the technology of nuclear fission and all others would have no clue what all that was about. Romans could've been the gods of those times, of course if they didn't kill themselves first, out of too much pride, which actually they did manage to do it in a spectacular way even without nuclear fission.

In our days nuclear fission is being thought in high schools, every nation on this planet knows about it, so is not a godly thing at all or is not considered so by our modern science.

If Paul knew about nuclear fission, but not because he understood it but because he read details about it, and tried to teach it, would he look like a crazy man to his generation? I do hope your answer was—yes-.

What if he knew about it and tried to hide it from others?

Jesus knew about resurrection, and so did Paul. Which one understood how it works and was able to apply it?

I think both of them did know but only Jesus was able to apply it. That's why Jesus was rather ironic at times about the one "who is coming after me", the one who has it but can't use it.

Both knew something else but both knew that only grace could lead to understanding.

Jesus was patient and did not boast about it. Paul admits that he did boast about the knowledge he had.

I was prepared to write a very well documented book about all this, rather a scholastic kind of book. But I feel that very few will care to know all those details. By the way, if they do, the details are still out there.

So I'll make it more fun. Fun-nyy is the heart of a multi billion dollar industry. Make fun of anything or everything and you found your niche filled with cash in this life. Be serious and try to teach things that are not easily understood and you got a good chance to end up poor and alone.

God makes fun of us at all times; if he would give up on fun he would kill us all.

Was looking forward to go and cite Great philosophers and the lame opinion about Paul. Now I would rather come up with "my own" opinion about him. "My own" is in quotation marks because nobody can have his/her own opinion. Every opinion we have, in a way or another was influenced by previous information.

I formed "my own opinion" about Paul, based on what Paul sort of said about himself. Maybe Galatians is the best letter to read in that regard.

Many details about the most influential scientists, was one of my main targets for this book. I'll stay with few remarks that could lead you to do your own research on their biography and scientific work. This book could become an encyclopedia if I would go for each linked example in science or religions.

So from the heavy-duty book I was planning to write I shift to lite and funny.

You know what's funny?

In a book I wrote right after 9-11took place I claimed that bin Laden will never be caught. At that time I thought I was crazy to claim such an obviously logic thing.

Since I know I have the fun in me I'll use it again.

The reason Paul looked rather crazy to many is because he was trying to hide or find a great deal of info.

Nevertheless, people saw what they saw and that wasn't easy to ignore. They saw Jesus' miracles and his resurrection and that wasn't easy to erase with letters.

Correct, Paul didn't try to say that the resurrection did not take place, but he did try to minimize the human condition Jesus kept talking about.

Paul never led people to think by themselves, he rather preached to them so they won't have the chance to form their own opinion. In fact Paul did advise them to have a united position, while they could not understand what was taking place.

That is exactly what every religion is trying to do, to unify people under the same set of ideas.

Even today if we would witness a resurrection event, most of us would not be able to understand it. If we can't understand it, how are we supposed to have a united position about it? Knowledge is a continuous struggle can't and shouldn't give a united position. Paul and the church tried to misinform and keep knowledge faraway from people.

Since I choose to start with Paul not with scientific events this might not seem very appealing to people of science, especially people of science that *believe* in evolution without any facts.

I am not religious. In fact I think every religion is corrupted and each carries just like Paul's gift, a well-boiled seed in all their rituals. Paul's curse is present in every religion after Jesus.

All religions use the name of God to control people. That's why the church is so political and people forget about God while praying in such church for their political and material needs. There is nothing to blame about greedy prayers. God can answer that, unless Paul was really crazy. But since he made it to the Great philosophers, maybe he wasn't as crazy as people of science would like to think.

If you ever read Paul's letters maybe you know they are not the same. To each gente/nation he writes a bit different. Main reason, being the fact that he knows their religions. And he also knows how to undermine their religion while using Jesus' name.

To no nation did he write what he wrote to Hebrews and Corinthians.

Each of his letters show his great knowledge, but only in Corinthians he matches one for one what he claimed all along. Of course, it is not complete, but has plenty of information for our times to understand it; Corinthians had no chance to understand it, but sure did sound out of ordinary to them.

The main idea is the resurrection in every religion. Islam would not exist without the resurrection, but the religion became so corrupted over time that claims there was no resurrection.

Without the resurrection, humanity would not even care to remember Jesus by now. Not even if the church would still be burning the sinners as in inquisition times.

Strangely enough, Vatican just released secret inquisition files after 700 years. Why was it necessary to hold on to that lie for so long? Did they try to cover up the way they got the power and control over people? Now might be too late to be honest is already 2007 A.D.

I'll cite that part of Paul's letter, because is too much to explain what he says without using his own words.

1 Corinthians 15) 20-50
"But Christ has indeed been raised from the dead, the first fruits of those who have fallen asleep. 21 For since death came through a man, the resurrection of the dead comes also through a man.

22 For as in Adam all die, so in Christ all will be made alive. 23 But each in his own turn: Christ, the firstfruits: then when he comes, those who belong to him. 24 Then the end will come, when he hands over the kingdom to God the Father after he has destroyed all dominion, authority and power. 25 For he must reign until he has put all his enemies under his feet. 26 The last enemy to be destroyed is death.

27 For he has put everything under his feet. Now when it says everything has been put under him, it is clear that that does not include God himself, who put everything under Christ. 28 When he has done this, then the Son himself will be made subject to him who put everything under him, so that God maybe all in all."

Rather clear that Jesus was asking to save humans and God is helping him to do so. But when Jesus' job is done after God puts all enemies to his feet, including death, Jesus has to turn this over to God.

"29 Now if there is no resurrection, what will those do who are baptized for the dead? If the dead are not raised at all, why are people baptized for

them? 30 And as for us, why do endanger ourselves every hour? 31 I die every day—I mean that, brothers just as sure as I glory over you in Christ Jesus our Lord. 32 If I fought wild beasts in Ephesus for merely human reasons, what have I gained? If the dead are not raised,

"Let us eat and drink, for tomorrow we die"

33 Don't be mislead ; Bad company corrupts good character 34 Come back to your senses as you ought and stop sinning; for there are some that are ignorant of God—I say this to your shame."

"*Do not be mislead; Bad company corrupts good character*" We could say that again and again and each time becomes more accurate than before. Yet, Paul knew how great his influence was over them especially while talking on this theme.

Resurrection was the theme that could have anybody in those times hold their breath while listening, but how much could they understand it's a different story.

"35 But someone may ask "How are the dead to be raised?."

With what kind of body will they come? 36 How foolish! What you sow does not come to life unless dies. 37 When you sow you did not plant the body that will be, but just the seed, perhaps of wheat or something else. 38 But gives it a body as he has determined, (is he talking genetics?) *and to each kind of seed gives his own body.*

39 All flesh is not the same. Men have one kind of flesh, animals have another, birds another and fish another. 40 There are also heavenly bodies and there are earthly bodies: but the splendor of heavenly bodies is one kind, and the splendor of earthly bodies is another. 41 The sun has one kind of splendor, the moon another and the stars another: and star differs from star in splendor. 42 So will it be with the resurrection of the dead. The body that is sown is perishable, it is raised imperishable, 43 it is sown in dishonor it is raised in glory; it is sown in weakness it is raised in power ; 44 it is sown a natural body, it is raised a spiritual body.

If there is a natural (physical) body, there is also a spiritual body. 45 So it is written: "The first man Adam became a living being ; the last Adam a life giving spirit. 46 The spiritual did not come first, but the natural (came first), and after that the spiritual. 47 The first man was of dust of the earth, the second man from heaven."

And if you continue to read the entire chapter 15 this idea becomes more intricate.

My question is;—How come Paul knows all this with such accuracy? Yet often he claims he does not know, and what he says is not from God but his own opinion. That is in fact where Paul confuses all those who read his letters.

In few cases he claims "I'll tell you a mystery or a holy secret". Yet, soon after he says that, he calms down and claims that what he said is only his own opinion.

How did Paul know that there is a physical body before a spiritual body?

Yes, he claims "it is written". But one has to admit, that what "it is written", in the Scriptures, sounds a bit different from the way Paul explains it. Is Paul correct?

Did Paul just interpret the Scriptures in the wrong way? Or he knew for a fact and so did Jesus about this kind of transformations of bodies?

Our modern science proves Paul correct about this kind of transformation. Energy can be converted in all kind of forms. But our modern science doesn't really know what energy IS. Nevertheless we can convert all kind of forms of energy to other forms. The main idea is to use a good efficiency of conversion.

Einstein claims that in fact matter is energy. He came up with that famous formula of his, which can't be really tested in physical reality, only by means of other rather valid formulas.

Lavoisier when he discovered the law of conservation, discovered just this principle, but wasn't able to go this far.

Our modern science still uses Lavoisier's law of conservation, but still can't understand the laws of transformation.

Paul on the other hand seems to understand which one has to be first in order for the second level to develop.

Paul knows that the "natural" body *has to be first*.

Our science knows only laws and principles for the natural/physical body but there is no knowledge about the spiritual body. In fact, science, would rather deny such possibility.

Could it be that in order for energy to exist, matter has to be its first form? We do not know about such succession yet.

Modern science doesn't accept easily the idea of secondary (spiritual) body, but admits that the physical body can influence very little if at all, the vegetative body.

One could claim that the vegetative body is still the natural body, which might be true, but in that case our science can't understand the components of the "natural" body and its mechanism yet.

Did Paul know about its mechanism and that's why he tried to perform miracles?

Our vegetative body controls the healing process in our body. Modern medicine will agree with that idea but can't use it in scientific way just yet.

Paul used that idea by means of "developed spiritual body". Well Jesus used it, but he did not explain the mechanism, or if he did, that's why Paul and the church "wasted" his written work. Some still claim they have Jesus written words. If that is true, by now I would be able to tell them apart from Paul's or any other writer on this planet.

That's a huge claim on my side, but if Jesus was indeed the teacher, he wrote astonishing things about human condition. Paul could never compare to Jesus' deeds, even if he exhibits more knowledge. Yet, according to me, Paul is the only one I know of, who mixes science and spirituality to a level we can't fully understand even today. I can imagine how much more Jesus knew about it, so would be easy to recognize Jesus behind the words. Jesus would go more like "watch and learn", while Paul goes more like "I know is true but I can't make it work yet".

How do I know that Paul had such knowledge of such text? He himself gives it away in few places.

For instance almost every time he cites the Scriptures he is very careful to which nation cites which part of the Scriptures.

As was rather obvious in the above letter, he kept citing Genesis and Jesus deeds in an unusual link.

"*Cor 15-47 The first man was of dust of the earth, the second man from heaven.*"

Nowhere in the OT or the NT is this link made in this fashion. Man of dust and man of heaven as unity.

Isaiah tried in few places to explain how Messiah will be, but still not this obvious.

Of course this is the same idea as natural body followed by spiritual body.

What is rather amazing about this expression is the fact that Paul talks in a very non spiritual way about the man of heaven. At some point he gives the feeling that he knows who this "man" of heaven is.

Here is another letter that has another amazing remark, which is found in the Scriptures. The way Paul presents it becomes a singular case in all his letters.

Hebrews 11 1-7 "Now faith is being sure of what we hope for and certain for what we do not see. 2 This is what the ancients were commended for. 3 By faith we understand that the universe was formed at God's command, so what is seen was not made out of what was visible. 4 By faith Abel offered God a better sacrifice than Cain did. By faith he was commanded as a righteous man, when God spoke well of his offerings. And by faith he still speaks, even though he is dead.

5 By faith Enoch was taken from this life, so that he did not experienced death; he could not be found, because God had taken him away. For before he was taken away, he was commended as one who pleased God. 6 And without faith is impossible to please God, because anyone who comes to him must believe that he exists and that he rewards those who earnestly seek him.

Almost everyone who went to church is familiar with ideas of these kind; Resurrection, immortality, faith followed by grace. Seems like a continuous transformation and by different means God tries to make people aware of this.

In this regard Jesus was the supreme model for humans.

He was born just like any other human on this planet and grew up, with few exceptions, almost as any other child. Or did he? The childhood years, which were not allowed in the bible, say he was a rather mean child.

After he became a young man, from 12 to 30 nobody dared to describe his looks or his life.

If one reads Isaiah 53, one might see a reason why Jesus was a solitary person. In Isaiah 53 Jesus is being described as a very non attractive person, so much so that one would have to turn his eyes away from him. On the other hand Isaiah wrote that text, about Jesus, about 800 years before Jesus.

Also is strange that about 500 years before Jesus, Plato described Socrates as the ugliest man he saw. Socrates was killed for his faith, without writing a word. Jesus was killed for his faith, without writing a word about it.

Paul doesn't talk much about Jesus or his childhood, nor do the disciples or those who wrote gospels about him. Jesus himself, according to the written gospels never talks about his own childhood or adolescent years, not even by mistake.

Jesus never said in the written gospels that; "I heard my father talking to my mother about God and how I was conceived without the husband of my mother". Actually a son should never hear about such things, even if God is the father. I could get even more ironic than this, but still won't help.

The action that was taken by the Council of Loadicea in Phrygia in 364 AD when a lot of books that today are in the bible were not accepted to be part of the bible so became the "lost books".

Among the banned books were;

The First and second book of Adam and Eve.

The secrets of Enoch,

Paul and Seneca,

Paul and Thecla,

Christ and Abgarus,

Infancy I&II,

Magnesians,

Philippians,

Philadelphians,

Eve,

Mary.

Many other banned books that never made it back into bible where they belong.

Some of these books did make it back to the bible over time, but others never did. Like infancy of Jesus never made it back to the Bible. Should Jesus' infancy be in the Bible at all? When comes to the New Testament would make a lot of sense to have Jesus' entire life in it. After all, according to "modern" Christianity Jesus is God himself.

"The Secrets of Enoch" never made it back to the Bible either. The "early church fathers" had the power to eventually change even the text in some of the books used in the Bible.

Maybe that's why the last sentence in revelation and therefore in the Bible claims;" *I warn everyone who hears the words of the prophecy of this book; If anyone adds anything to them God will add to him the plagues described in this book. And if anyone takes words away from this book of prophecy, God will take away from him his share in the tree of life and in the holy city, which were described in the book.*"

The Secrets of Enoch, according to Tertulian, was in fact the Holy Scripture before the Christian era begun. Should that book be added to the Bible? Should the book of Abraham be added to the Bible? In my opinion, if these books are not worthy to be in the Bible no other books should be.

Paul and Seneca did discuss topics that were not allowed in the church. Both of them were philosophers.

Often, Paul in his letters reflects the philosophy of his times. Ideas that are obvious as Aristotle's influence can be easily spotted in Paul's letters. That could be the topic of a different book, which was probably already written, but I didn't have the chance to read it.

Not even Paul, who eventually was the most spiritual philosopher, was able to have totally original ideas.

His ideas are still the most influential ideas in the structure of our society. The church borrowed his ideas over Enoch's Book of Secrets.

How did Paul manage to influence the church to such extent?

No other philosopher made it to the bible. Paul made it to both.

Socrates lost his life by being a firm believer but didn't make it to the bible. In fact his ideas were discarded by church. One of Socrates main idea was to educate women as well as men if not the society will have a "missing arm". According to him the integrity of the society will be lacking if women are kept from education.

Meanwhile Aristotle claims that women are but the soil in a society, they don't even participate to human seed, only males have human seed.

After 2000 + years we know that Aristotle was wrong and Socrates was right. Thanks to modern science we know that human seed is made of both genders. Even more, the women' side of human "seed" carries the "genetic repair kit" as well, which is known as the mitochondrial DNA.

Paul borrows Aristotle's ideas and used them in his letters and gets them in the Bible.

This is the part that amazes me about Paul.

Jesus himself didn't make it to the Bible and Great philosophers the way Paul did.

At the same time Paul was the one who found or used Jesus' writings, if Jesus ever wrote.

Besides Jesus' writings there was plenty of ancient information by that time.

Paul's letters, if indeed his letters weren't altered by the early church fathers, do have a lot of contradictions.

Someone like Paul would not contradict himself in such an obvious manner.

At the same time the Bible is set up in order to contradict and confuse those who won't take the time to read it carefully.

In times when the Bible was not allowed to be read in one's native language, but was forced only in Latin, meant only a lot of empty words to most.

In those times people didn't read by themselves because in a way they were waiting or even forced to accept the priest's interpretation of the text plus very few were able to read to begin with. In fact to this day many are being forced into accepting everything catechism says. Well, today is rather one's choice to accept and follow with no questions asked. Today most people can read, most people have choices to think or not to think.

There were times when one could be killed for having the holy text; people weren't allowed to own the book, so became the holy book, which became over time, rather a fairy tail than real life events.

Even more so, over time, Jesus was made God, so indeed the resurrection became separated from the human condition. The man of dust and the man of heaven were separated by church.

The more pressure the church leaders put on people the less they cared about God. That was in fact the intention, so the church could control the people.

The church had all the interest to do this kind of move. The church leaders were in control of the "lost books" and much more. Inquisition was a necessity in those times to keep people away from truth about Jesus' resurrection and from his teachings about human nature. Paul, on the other hand, writes about "holy mysteries", showing the link between man of dust and man of heaven. For that very idea, Inquisition would burn to death the "heretics" who believed it. Inquisition did force the belief that Jesus is God.

Many never got the chance to read or hear about Jesus' childhood, or the Secrets of Enoch or many other books that would or could destroy the image the church was trying to create about Jesus and resurrection.

If Jesus was God, than indeed resurrection is not for humans.

Yet, Paul claims in Corinthians 15 for the entire chapter that resurrection is part of human condition not a God condition.

God is not made of dust and God of heaven. The idea of God of dust and God of heaven wasn't part of Paul's letters.

Hebrews 11-5 clearly says that Paul firmly believes that Enoch by faith did not experience death. Therefore Enoch is still alive.

Only a mortal can become immortal, according to Paul. That idea was said long before Jesus, and Paul knew about it. If all these ideas are correct, they are in fact logic too.

Why did I choose Paul to start this book?

Because it is easier to start with the one who connects not only in theory the "real" life with the "imaginary" life.

It's easier to use the one who made it into the Bible even over Jesus' infancy book, and made it to the Great philosophers at the same time.

But what is the most intriguing to me, is the fact, that Paul's ideas still hold the secret to future science.

From where did he get those ideas?

Was he really the one who had the Holy Grail or was he the one who knew of its existence and tried to find it allover the place?

No human mind can come up with ideas as Paul's by simple chance. Paul had to know about this kind of transformation of body. He had to know exactly what a spiritual body is, and why has to follow a physical body.

Everything he says was indeed "given" to him. When it wasn't given he claims it is his own opinion.

But let's leave Paul for awhile and try to move on to science. Or to what we like to call "modern science".

I saw the other day some show about pyramids. Was funny to see how hard some really think they could build pyramids these days. In the end they built a pyramid as big as an outhouse, and couldn't stop bragging of how easy it is to actually build pyramids.

Was funny to see such confidence after the outhouse was done. If you'll read this book to the end you'll smile too to the idea that we can build pyramids.

Everything is possible with the right "tools".

If you did hear the story of Ed and his Coral castle, you might know what I know.

In most cases that is how modern science treats everything.

One comes up with a big bang idea, and the universe gets into existence. One builds an outhouse of large stones, and the presence of pyramids in the oddest places on this planet and their significance becomes nonsense.

Human nature is really funny that way.

The role of science is to imagine, and diminish the truth.

Many imaginative people supported the idea of evolution. Some even claimed that species came into existence by evolving from each other, due to the adaptation or best fit to environment "laws".

That's as funny as it gets. Is a dog best fit to environment than a human? A dog has four legs to run, has really sensitive smell, self protecting coat, and has the ability to sense danger and much more compared to humans. To loose all that adaptation to environment and still call the idea of best fit sane is just funny.

In fact all the "inferior" species are better adapted than humans are to the environment on this plant. None of the specie needs to use fire in order to maintain body temperature and still survive on this planet. Does that mean that in fact involution took place when humans lost their ability to run as fast as a dog, to fly or climb trees or smell, or even be able to read the "map" as a bird does?

Back in 1800s (please note the time frame), when Mendel came up with segregation of characters, evolution theory was looking for more support.

The segregation ratio came as thunder bolt, to destroy the theory. But couldn't, because the then "modern science" was so involved in proving evolution correct, that denied the ratio as correct for more than 50 years.

In fifty years, the social structure changed. By early 1900s when evolution seemed to have too many holes in it, the ratio of segregation presented by Mendel started to gain interest.

The reason why science, back then was looking into it, was simply because they hoped to connect the segregation ideas with evolution theory.

As one might expect, the most prominent scientists at that time, claim that the segregation ratio is the punch of death for evolution theory.

If I recall correctly Thomas Morgan said that, after he run few extensive studies on Drosophila melanogaster or the fruit fly.

The most prominent scientists even at this time completely neglected that idea, and still tried to find different ways to prove evolution theory as being correct.

To this day, the evolution theory was not proven. Even worse is the fact, that every auxiliary theory to prove evolution as fact failed, based on recent scientific discoveries.

The world of science still debates this topic.
This "scientific" debate might go on for another few generations.

The most amazing aspect is the way Mendel discovered his laws, yet nobody believed him for decades.

On his death bed Mendel said in full confidence (I'll paraphrase for now, but I'll look it up for later, his last words)
"" what I did present is correct, even if not recognized by science, will be recognized by future science"".
Sounds so familiar with what Newton claimed about his own discoveries. ""my work is not for this generation"". In fact Newton published his work only after few of his scientific adversaries died. Newton's ideas, if published while his adversaries were still controlling the "scientific arena", would've been lost in their adaptations to then scientific level.

Let's go back to Mendel for a while.
Is known he was a monk. He also knew about the "modern science" of his times. Evolution theory was something he did not expect to hear.
Being a very devoted believer, he started his own work, or was it his own?
Many will say, yes, he did work in his garden playing around out of nowhere with peas. Not only did he see that peas had different shapes, but he did hybridized them for a purpose, which wasn't for eating.

He knew if he does that he could be expelled from monastery. According to some he was expelled in the end.

So he dies rejected by scientific world and rejected by his church. Who would do such thing, if he didn't know more than he could explain to his generation?

Why would Mendel go against the "modern science" of his times, while disobeying God's word as well?

Did he know something with such certainty that would give him such courage to prove modern science of his times wrong, and get in really bad terms with "God" for doing so?

These are some rhetoric questions in a way.

What I found unbelievable in Mendel's work, is not his attitude towards church or modern science of his times.

Not even his discoveries were that impressive to me, but his method and most of all his certainty.

To claim, after decades of being rejected by his times scientific minds, on your deathbed, that your discoveries are correct and future science will prove it. To be 100% correct not only about the exact ratio, but also about the fact that future science did, indeed, prove his laws correct for any character. All that is still good, but the fact that modern science still can't understand to this day why the ratio forms in the first place is intriguing.

That's the aspect that kept me awake for a while.

First let me talk about methods.

In science/scientific research, the methods used are error and trial, which means; if this way doesn't work, let's try another. Let's say a chemical concentration doesn't work for a specific purpose/effect. That concentration is eliminated from the trial and replace with a different chemical concentration. If at some point a certain concentration will induce the effect that is search for, all concentrations that did not will be eliminated. Let's say 1000 ppm is the optimum concentration. At this concentration the productivity of tomato plant was at its peak. Below or above this concentration the productivity was lower or the same. This result was achieved based on trial and error.

A method is usually developed on pre-existent research and uses previous patterns, or part of previous patterns. A method is never a "stand alone" method as in the case of Mendel's law of segregation.

He simply did something that clearly wasn't an accident. He wasn't even allowed to conduct experiments in the place he was at. He doesn't follow previous patterns from his time "modern science".

Mendel hits the nail on the head without trial and error, without copying pre-existent patterns. He doesn't fall in the trap evolution theory had for all scientists in his time. All other scientists in his time were trying to adapt their theories according to evolution theory. The same is true today. The evolution theory was the giant every scientist aspired towards proving it. Same aspirations are true today.

He stands alone in all aspects. Nobody can do that without an outside source of information.

If you ask yourself what was his source and you care to know, you'll find out few pages later.

This is not because I want to create any suspense moments. This book is not about such ideas.

If I tell the source now I can't continue to explain why Newton or Mendeleyev did the same.

Out of all these people of science, we know by now, only Newton did tell the truth about his source.

Many ancient books were lost, but Newton claims he had the chance to buy some ancient books. He found out in those books about the laws of gravity, but he didn't find out what gravity is. Is that possible?

That sounds to me more like Paul talking about spiritual body, and the way the transformation takes place, but has no idea what spiritual body is. Is that possible?
To know about its existence, its law of transformation and not know what is it?

Yes it's possible. All modern science does and did is just that.
Genetics uses the law of segregation, but doesn't know what is it or why does it take place the way it does.
Electromagnetism and related electricity effects are known, but not understood.

This is the great enigma of modern science. All inventions, all discoveries, everything that human mind can imagine is based on phenomenon we can't understand. Nor why do they exist nor what they really are.
If Paul knew that the spiritual body follows the natural body, didn't have to know what spiritual body is.
Jesus on the other hand knew what spiritual body was, that's why he went along with the process of transformation.

Do you recall in Da Vinci's code, Newton's name was mentioned few times. Even though, the book was talking about him, did not mention about science or anything related to his scientific career.

Intriguing, since both, Da Vinci and Newton were pioneers in what they did "discover". Also few others were pioneers around the beginning of 19th century, but each in different fields. Maybe Newton was the only one who had most information in one package. If Newton had Mendeleyev's information he would've been able to carry out his "alchemy" project.

Mendeleyev knew the laws on which he did develop the table of chemical elements, and left open the spaces where there were no known elements for. He was very sure of the existence of missing elements.

Our modern science proves he was right. Still is not fully understood how the laws Mendeleyev used to put together the table really work. He himself did not know.

Just like Newton did not understand gravity, but did "discover" its laws.

This time frame, 1800 s, boosted or started the "real modern science" era. All of a sudden around 1800 all the major scientific laws were discovered. All of them were rejected by then "modern science" as illogic or incorrect. Mendel's law of segregation was accepted by science after 50 years or so.

There are few others who had impacted our modern science in ways that weren't accepted by the scientific arena of their times.

I won't go in their biography or give too many details about them, since will be too much information that already exists and can be found easily.

In America, Edison, seems to be one of them.

Did Edison have any parts of the Holy Grail?

Maybe he did, maybe he didn't, but most around him admit he wasn't a "brilliant" man by their standards.

Edison made crucial discoveries without "classic research methods". His work was much targeted. His inventions had no scientific basis at that time.

In fact the world of science rejected him as scientist to this day. At best he was an uneducated "inventor".

Which means he had no clue what kind of phenomenon was using, but he "got wind" of how to use the principles.

Phenomenon is still unknown to this day.

Strange how our modern science can't penetrate the shell of phenomenon but can use its laws.

We measure the force of gravity, but don't know what it is or what causes it.

We use electricity, and many laws that apply to it, but can't understand the phenomenon called "electrum".

We can convert types of energy, from caloric to nuclear, or the other way around, but do not know what energy really is.

We, as in our modern science, know everything there is to know but we do not know why it is.

Science prefers to call these questions rhetoric.

If Mendel would've had all the information about gravity as Newton did and chemical structure as Mendeleyev did, he would've been the king of the world. Maybe a mad king, sine nobody in his time could possibly know what he is talking about.

He would've gained tremendous power with no way of being controlled by usual means.

Almost easy to understand why there was a need of a grail/network to release such potent social poison as knowledge is.

If anybody can deny that the parents of our modern science were most "active" only in a certain time frame, which was around the beginning of 19th century than we have a chance to believe that human mind is really not able to come up by its own ability to realistic scientific conclusions. Too many generations lived in the dark.

I know what I just said could hurt the pride of the most influential scientists of our modern times. Should we let the pride and arrogance to destroy the information that was given to humanity long before there were pyramids on this planet?

That's the beauty of religions. Every religion corrupted the truth they use due to pride and greed, but didn't think that will come back to them like a powerful boomerang.

Christianity made Jesus God. Islam would not exist without Jesus resurrection, yet diminishes his resurrection.

Atheism claims there is no God; yet, doesn't have any explanation of any of the above phenomenon. Actually according to some atheists there is an option that A god exists, which can only confuse the entire ideology even more, and that's also a sign of corruption.

Of course I am talking about ideological corruption.

Most people love to add that "extra" to their ideology. To make it sound unique, special, more original than any other religion/ideology out there.

That's exactly where the destruction of truth begins in science as in any ideology/religion.

Science doesn't try to look into common aspects in different fields, but tries to enhance to extreme one field.

Instead of common ground for scientific fields, there are great divisions among fields of science.

That's the modern emphasis in science and its way to self-destruction.

Yes, it's true that one has to understand each area in order to reach common ground. Maybe in the future that will be the approach, but pride in the scientific arena will play a huge role in order for that to happen before will be too late.

It's easy to imagine that next generation will be so confused by modern science that will ask "time out" to re-group.

What took place back in the 1800s is not happening today. Our science still stands on those discoveries.

If anybody would come with a rather solid theory to support evolution for decades to come that would be the main target of our modern science. Also that would become the scientific doctrine.

To see in America such a lenient scientific pattern is rather discouraging. That's why I choose to write this book more on the funny side, than on the scholastic side.

Plus if I choose to write a scholastic book I couldn't deal with so much information alone in the short time I have left.

So let's go back to funny side of the story.

Did you hear lately that church and state can't co-exist?

No wonder they can't co-exist. When did they in fact?

Oh there was that one time in human history, but turn out really funny. The greatest king ended up eating grass for seven years. After that he came back to "normal" life and was considered crazy. Yet, this is the king that changed human history forever. Of course, that's because things didn't work out the way he planned.

By now you might know who that king was, if you don't I won't tell you yet.

Why not? Maybe because I can do it, I choose to control what I want to let you know, so I can mislead you if I want to.

Oh you got me now. This story is way too popular, so you knew it.

Darn, I can't keep a secret like this for too long. Daniel told you all about Nebuchadnezzar.

This king had it all. He was the king of kings. He was so powerful that he wanted to bring the entire world to one religion. He made a huge god of gold and forced all the executive commanders and leaders from allover his empire to accept this new god/idol of gold as their new god. Once they did that they would have to go back home in their particular areas of the empire and enforced there. So in the end the entire empire will have only one god of gold and one religion; to worship that man made god/idol.

He had a huge furnace to enforce that system of believes. Believe in the new god of gold, accept him as your only god, unite under this god or die in the furnace. All of them accepted the new god, with the exception of three. Of course these three they believed in the God of Abraham. God of Abraham was the most potent God, which is why the king of kings was afraid that things would go back to this potent God.

Funny thing happened, after the king of kings "allows" the three to get out from the burning furnace since he the king saw a "son of God" with the three in the furnace.

He, the great king of all, gives a decree; "that all those who won't believe from now on in the God of Abraham, should be killed and their house turned into ruble".

I find this story very funny.

God of Abraham playing the king of kings this way. To force him in matter of minutes to change his mind about God, in front of all his influential representatives from allover his empire. His fabricated gold idol turned to rubble right there in front of all his commanders and regional representatives. No wonder he went crazy for seven years after his pride was crashed in such a spectacular way.

Why did I tell you this story? Maybe because a story like this can be used as a reminder of how potent and damaging human pride can be?

Well of course he ends up eating grass for seven years and then he came back and declared that God Almighty is the true God. By that time, nobody believed him. And he was declared insane.

He was the king who introduced the first monotheistic law by his induced decree. Of course, things did not go as planned.

Why would a king like him want to have everyone believe in a god of gold, made by human hands?

Maybe for the same reason why our modern science tries to prove evolution correct without success?

In fact modern science proves evolution or rather speciation impossible.

Now, since I did spread your thoughts over time and over fields of science and ideologies we should try to focus for a moment.

We could focus on human pride. Human pride is the leading cause of our wrong decisions. Pride doesn't allow us to be honest to ourselves. Pride requires power over your own kind and leads to self-destruction by rejecting our own inner being.

Out of pride people have ideologies they love to belong to. We love to defend our ideology even if we do not understand it.

This is true for all ideologies. I am sure that the god made of gold, would've been the god of that empire for centuries to come. Many would've killed for it, out of fear or simply out of devotion. The need to belong to a strong ideology is the need required by our pride.

As soon as one becomes a leader of some sort his persona is altered by the "I am a good Muslim or Christian or atheist."(Strange the Microsoft word did not require capital A for A theist, so I'll leave it to that)

Root of every war or every stagnating scientific theory is pride.

That's why science makes huge steps only during war times.

Out of pride we loose the ability to listen or the desire to understand other ideologies. We love to get fixated on one doctrine and defend that forever, if possible.

Atheism is the ideology that has as basis pride. Atheism helps people think very highly of themselves.

By being so focused on oneself people forget to look into deeper meaning than their own achievement being superior to all others. My house is taller than yours, kind of mentality. My car is more expensive than yours; therefore I am superior to you.

On the other hand, in an ironic way, teaches modesty, which becomes false due to the very fact of being "modestly superior".

Maybe this is a difficult point to understand so an example could be of help. If you have one please use it.

An educated atheist will always want to stay in that comfort zone of intelligent environment that gives confidence and self-realization. As soon as that aspect is under scrutiny the pride will tend to defend it under any circumstances. You might say that is the case with any given ideology, which is true. In the case of atheism, is more critical to defend it, since there is nothing else to rely on.

Any other ideology has an external force or intelligence to rely on. That force is not supposed to be fully understood in order to be accepted. Nevertheless many become very proud of being part of such ideology.

Just like gravity; *Exists* even if we can't understand it. We use its laws to a certain extent and we have to be content for now with what we got.

In the case of atheism, is no need to be understood either, but even so there is no need to rely on what we can't understand. This seems logic at first glance.

The fact that our mind can't understand should be the aspect to consider most dangerous. We can't understand, yet we choose to defend. Nothing could be more damaging to our integrity.

Is more like choosing to defend a violator over a non-violator. I can't understand the non-violator, but the violator feels closer to home and it's easier to understand.

Like choosing to kill Jesus over a criminal, just because is easier to feel for a criminal, than to try to understand a confusing idea that undermines the ideology you try to defend.

So by choosing to kill Jesus one already chooses to protect their ideology unaltered. The criminal has no ability to affect the ideology. At best the criminal will embrace the ideology that saved his life.

Wars are mostly triggered by greed, but ideology is used to control the people.

By telling Christians that Muslims could affect their ideology will trigger the need to defend Christianity. The vice versa is also true.

The benefit, if one can call it that, will be taken by very few who couldn't care less about either ideology.

The Pope selling indulgences came to mind. The Pope just wanted to make more money, he couldn't care less if the sinners are forgiven or not.

Vatican had the power to do that, and the ones in power used it. Luther was strong enough to change that for a while.

Due to pride, ideologies are being embraced. Being part of a large group makes one think that its members can't be all wrong so they must be correct. History proves that the larger the group the less they will care about being right/correct. The desire will be to enforce the rules of their ideology and "unite" people under the same ideology. Does that sound familiar? From Nebuchadnezzar to our modern science this pattern remains unaltered.

Yet, when Nebuchadnezzar came back to "sanity" and told the truth he did discover, he was considered crazy, rather than listen to him. What would any of us do in such a case? Would anybody listen to the mad man who used to eat grass for seven years? I don't think I would, especially if I was part of an influential powerful ideology. Who could be correct the mad man or my all-powerful religion?

Indeed; "The wrong and strong is all-powerful and favored by most". I did hear former president Clinton saying that on TV few years back.

Being part of the wrong and strong is the dream of every proud persona. By being able to say, "I am part of this great ideology" one feels in the comfort zone and also all his/her faults are excused by the "being part of". This kind of mentality sounds more like feeling better to lean on a two by four in quick sand, than searching for solid ground.

That was indeed too condensed. I used the two by four as the lean on, because most people love to hate or fight other ideologies. The two by four is the symbol of fighting over nothing, to me.

Now since we consumed part of the pride issue, might be easier to explain why people of science feel so proud.

As most of us know, when one thinks that knows more than another that gives us a feeling of superiority. That should be acceptable, after all one cared to learn more than another. The critical aspect is; does this person of science know the correct information in his her field of study? Does that PhD really entitles one to reject everything else?

One can only imagine that all those people of science who rejected Mendel's law of segregation and did not care to verify what he claimed were simply misinformed in their field of study. Nothing but pride stopped them from verifying Mendel's laws.

Fisher, who was an influential persona of science in those times tried to get rid of Mendel and his law, since indeed this "logically illogic" or counterintuitive way was undermining the then modern evolution theory.

Agreeing with Mendel's law (s) as being correct would mean to stop and think about evolution and its selection.

Why would segregate in a 3 to 1 ratio for any given character right after a great "selection" of the fittest did take place?

That would've been impossible to explain in a logic way. The fittest "gave up" the great selection that took place, maybe for millions of years? Like that wasn't enough, would've been even harder to explain the constant ratio for any character. That couldn't be accepted as logic in the logic of evolution.

Maybe that's why Thomas Morgan trying to support evolution runs few experiments on few generations of fruit flies. The outcome was exactly what Mendel claimed. Every character's segregation took place in a 3 to 1 ratio. Morgan was wondering why that was, but could not explain it. His final conclusion was exactly the opposite of what he was hoping to find. His results, as he puts it, lead one to think that segregation of characters in a constant ratio is "the punch of death for evolution theory". Unlike many others, Morgan was honest about his work, and he did not try to support any scientific doctrine. Just like religions, science used Morgan's work, but not his ideas, as scientific.

Been wondering for a while now, how come that such observations are not being thought in schools as accurate?

This was proven by means of experiments and is being proven today and used to induce "mutations". The word that is dedicated by now, should be revised by our modern science, what we use to call mutation is not really a mutation.

I won't debate this now, since could get too scientific for the purpose of this book. But a mutation that segregates how can it be a mutation?

For sure plenty of arguments can be claimed on this observation, but will take a while for biology to accept that what is being called mutation doesn't really mean what our science claims to be, since not only that is reversible due to environmental factors, but also segregates.

If there was a somatic mutation, like spotted foliage, which is transmissible only by means of vegetative cloning, that mutation can be reversible in a different pH, or different temps.

This is not new, yet, biology still calls them mutations. A reversible mutation simply means that a character that already existed in that genotype is manifested. Therefore can't be a mutation as is being claimed, is just a simple activation or lost activation of some function of a gene.

That's why most mutations will return to the "WT" (wild type) or original genotype in a less stressful environment.

At this point I could go scientific for a moment. In the fifth edition of "Genetic Analysis" mutation is defined as: "*Mutation is the process whereby genes change from one allelic form to another*". This definition, according to my old school of thought, doesn't have the correct structure; first because uses the word that is being defined in its own definition and second; if there is only a change from one form to another implies that the ability pre existed and nothing new or different was added. So the very core of modern evolution rests on mutation? I would rethink this definition in such a way that there is not allowed anything that did pre exist to be considered as mutation. That would indeed lead to new species and the segregation ratio would be impossible. There would be no return to WT but indeed evolving into new species. But, again, as some reasonable scientist would say; "this doesn't happen in nature".

Primarily Thomas Morgan used these mutation ideas, in a time when there was no other understanding of what really happen in the genome.

For instance the Galapagos "incident", with short wings considered adaptation to the environment, is in fact a stress triggered modification. This character can regain the original wing shape if the stress is being removed. That simply means that all specie have a certain range of adaptation to stressful environment and when the stress is over the "original" shape is regained.

Like, let's say, a fly that has the option to open an umbrella kind of wing when is raining, if that specie lives in the desert that character will be extremely reduced. The wing might be missing or if the genotype had a flexible gene about this wing, the wing might be changed to a shade providing umbrella. If this particular specie will be switched from the desert condition to the rain forest environment will have the ability to go back to its "original" or its WT form.

Therefore the "mutation" is a pre-existing character, just as the definition of mutation claims. This kind of mutation did not appear to be an out of range mutation for that specie' or genotype. A flying dog is rather out of the genotype's range.

Modern science still needs to accept this obvious truth. Morgan had no way of knowing about active genes or inactive ones. The observations he made were extremely valid for his time, but are incorrect for our times. Our modern science still preys on them. Trying to use the beaks of some specie as interspecies potential link is not the best option to find the truth about genotype limitations or flexibility.

Hopefully one day science will admit to this truth, which was present all along, but Morgan in the hope to affiliate himself to the big evolutionary movement of his time presented as potential induced mutation due to environment not as pre-existing inactive genes. To this day that very idea is still being thought in schools. If this adaptation to environment would be carefully analyzed we could come up with the obvious observable fact. Every genotype is stabile. Therefore its limits are known. In within these limits the flexibility to deal with stressful environmental factors is limited but present.

Again comes to my mind my excitement from the times I was embracing evolution as correct.

One day I read that same specie can "adapt" to different condition of environment, when the environment is optimum that "adaptation" is lost or returns to the WT.

One of my friends, an avid collector of "mutants", came home from some 3000 meters altitude with a variety of "miniature salix" or wiping willow. Very excited with her new found "baby" she carefully planted the salix in the best spot in her yard. Few years later the miniature became a "healthy" wiping willow as tall as her house.

Her excitement about the miniature "mutant" Salix was gone and had no explanation to what just happen to the miniature side of Salix, since in the place she found it, the left behind ones were still miniatures.

To me is rather clear that the Salix seeds flew too high up and started their life at a stressful altitude. If they had no ability to adapt in their very first generation to that stress they would've perish and there was no chance to find any miniature "mutant" Salix plants.

The ability to "adapt" in the first generation is not really a mutation, but a pre-existing condition. As soon as the plant found optimum conditions, grew right up in the same generation. Would one call that "reversible mutation" if took place in the same generation? I don't think that would be correct.

Certain aspects that took place for most of the characters he did study, often confused Morgan. In trying to align himself with the scientific movement of his time, he explains this flexibility in genotype as mutations that could lead to interspecies links. Those links were never observed by anybody so far; yet, science is still looking in the wrong places for them.

If an "adaptation" to environment is pre-existing that individual should be able to deal with the environmental stress factors that will be exposed to. That "adaptation" is limited and what is disturbing is the fact that could be reversible in the same generation.

Out of pride this obvious aspect will not be embraced by science in the near future.

Every ideology is corrupted the same way. People who got the control over certain aspects in a religion or in science won't accept the obvious if that means to undermine "well established" patterns of thinking. Most scientific projects would have to be cancelled or to accept a shift in their targets.

At the same time that could lead to social turmoil, since most people do not question what are being told in school or church. A grade is more important than a question that "looks totally stupid".

Hope the point I wanted to make came rather clear, but I am afraid that the obstacle to understand it is the knowledge most people already have as their treasure.

Very few will accept to follow my simple logic. In most cases I rather encountered a rebellious "atheistic" opinion.

For sure I do not expect to be understood with the simple examples I gave so far, while most of us went to school for years and we were graded A pluses for our impetuous knowledge about how Homo erectus came to be. Don't want to be ironic, but all there is about this idea is just that primitive drawing I saw 30 or 40 years ago. There are few levels from monkey to Homo erectus. A simple drawing that has no scientific basis and is but an assumption forced on people's minds.

Everybody in the world of science knows that, yet nobody dares to stand up and face that obvious fact.

The reason is that most do not care. Those few who might care have a family to take care of. And those very few who would like to "divulge" this obvious truth do realize the crisis they might induce for the entire world of science.

So when an assumption becomes extremely obsolete will be replaced in such a way that won't affect very deep the pillar science is hanging on. From Morgan to modern research that pillar is the theory of evolution. What an imaginary pillar this theory really is! Yet, thanks to that pillar our science did advance. I find that extremely ironic. Only after Darwin's theory was accepted as science, things really start rolling.

Mendel came up with his discoveries while the theory of evolution was gaining momentum. Nobody cared about Mendel's laws. He wasn't really the PhD of any field of science, so his work got tossed back for decades. In his time, as well as in our times, everything had to be explained by science trough the filter of evolution.

Mendel's laws were obsolete for that theory, but are standing still for our times, just as he predicted on his deathbed.

Evolution is still getting punches, but is changing its face every time it gets hit. The theory Darwin presented is obsolete for our modern science. The theory changed so much in the last three decades that nobody would recognize it if compared to its original, yet is still under the same name. When evolutionary "Botox" is not enough to keep it young and fresh, true face lifts have to come to protect it.

This effect reminds me of every religion. Every religion tries to adapt itself to modern times, therefore corrupts the truth was built on because the leaders never cared to understand it.

Religions accepting the theory of evolution as God's plan were a novelty few decades back. Now almost every religion embraced the corruption science holds on to.

Science claims Darwin's theory is obsolete; the church is still few decades behind, claiming that it is correct and that God made it that way.

Often the church did just that in order to keep the members happy and in tone with their educated times. Same goes true for the flat Earth theory. At first was in fact a scientific theory which later on was embraced by the church leaders.

People of science did join the church since science wasn't taboo anymore. Many tried to obtain power using their religious congregations. God was only the display on the door.

Power and pride corrupts everything humans will touch for as long as the priority is pride. Be the winner at any price is the slogan of any campaign. Every religion will try to influence as many people as possible even if they have to mislead them.

Islam will kill those who want to depart from it. Yet, Islam doesn't care to know the truth about the God of Abraham they are praying to five times a day. Christianity did destroy as much knowledge as they could about God. Books and libraries were burn to the ground. Those who dared to reveal any information found in ancient books were killed. Galileo is just a lucky incident. Leonardo Da Vinci had a lot of talent the church wanted to use that's why he escaped. We know by now that Da Vinci was a pioneer even in aviation. He also made the first fire gun. Somehow he would try not to make them work right because he saw the great damage of fire arms to humanity.

All these affirmations are well known. Entire world knows how science was oppressed by church, especially the papal power. Yet this is what God said right in the Bible Vatican was using to mislead the people;

Jeremiah 6-16 "This is what the Lord says;" Stand at the cross roads and look; ask for the ancient paths, ask where the good way is and walk in it, and you will find rest for your souls. But you said; "we will not walk in it"

To this day humanity acts the same way towards the knowledge God gave us and ask us to walk in it for our own good.

What else, but pride, could stop humanity not to walk on the path humanity was destined to walk on?

If I was told I have to walk a certain road I am sure I would rebel. I want the freedom to choose my road even if the destined road is the best possible road for me. The more we hear that God wants us to walk a certain path the more we say "we will not walk in it". Like a child that walks towards fire just because doesn't like to obey the command of the parent who keeps saying; "don't go near fire".

That stubborn reaction is common to all of us. As soon as we have some knowledge, even a child that has the knowledge of fire will try not to do what was told to do if they see no immediate reward in it.

If the reward is in front of our eyes we will walk towards it out of greed and pride. To have it and be the first who gets it.

If there is no reward even if the path is the best for us would have no value to us, since doesn't serve our pride.

Once we learn that our pride is but vanity, that vanity doesn't make any sense at all, our mentality will change. Only then can we learn to make the right decisions that affect our human nature. When pride and stubbornness is gone we can walk on the right path and not even choose it over any other path but admire its beauty. Only then we can appreciate how much better and what a great feature human condition really has.

The only problem is, when one chooses to walk the right path out of greed. To be greedy to follow the path, just like those who built the Bab-el tower. They wanted to reach to God in heaven, which was the right direction, but not the way their human condition was supposed to "walk". Many religions lead people to such "bab-el desire".

Islam is probably the closer to "bab-el desire" in our times. But Christianity went through such stage too.

Christianity did make a mortal, God. This choice has in fact two sharp edges. One is that God is a mortal just like us. The other is to keep the knowledge Jesus thought, away from people. Neither is true or correct, but the greed of some, mislead many.

This is only part of the great damage pride can do to us.

Anything we think or we gain usually makes us proud. A luxurious car feeds our vanity. A PhD usually fills us with pride, even if we know less than we knew before we had that degree.

How could that be possible, one might ask? By now I think is very easy to be mislead by science. If one goes to school for 5 years just to get out claiming there is nothing else smarter than he is. That in itself proves one knew more before he or she went to school. At that time he

knew there are smarter ones than he is. That was correct even after he
did finish school.

Well this was only one ironic example.

When God claims (too bad I can't find the text right now, some-
where in Jeremiah) that; "I destroy my people because of their lack of
knowledge", one would see this as a complete contradiction for everything religions stand for. Also God is suggesting to go to the crossroads
(which means go to places where people of all cultures go and cross
each other ways) to find out about the ancient ways. That simply sug-
gests to me that ancient knowledge was lost or changed in an inaccurate
way.

Would be a true miracle to have so many religions based on the same
book, that claim the same God of Abraham being their God and all of
them to be accurate.

What if none of these religions are accurate the way they interpret
the God of Abraham?

What if all these religions just fight each other without having true
knowledge? Could that be the reason why God gives advice even to the
Israelites to go at the crossroads and meet those who know more about
their true God? I said even to the Israelites, because they had Moses,
they had generations after generations that dealt with the God of Abraham in some form.

Why does God care to give us knowledge and we tend to choose to
be ignorant about it?

The most disturbing aspect is the fact that all religions that claim the
God of Abraham to be their God do not choose to go to the "cross-

roads" and search for truth. That in fact is the very taboo of every religion.

Indeed that would be difficult to do, since every religion claims that all other religions are wrong and inferior. Of course that boosts people pride to be part of the "correct" flawless religion.

"My religion is better than yours" mentality won't send anybody to the crossroads to ask for more.

"My religion is the perfect superior religion and whoever touches it will die" is the result of pride. That aspect is exactly what every religion that wants to dominate and control people will instill. Once a human mind thinks his or her religion is superior to all others there is no need to use external forces to keep one strong in their religion which is their faith, but not the search for truth and God.

But if a religion will go at the crossroads, ask for truth and be uncertain of what to really think will fall apart as religion and people won't love to be part of it since is not boosting their pride. All vanity and superiority is lost. Plus one finds him/herself on the crossroads listening to strangers whom they do not know or do not appreciate them. How is one supposed to listen to a hungry man at the crossroads? The prophets were killed because nobody cared to listen to them. What they said was usually damaging to "the all—powerful religion" people loved to belong to.

Today many try to exploit their ability to be a prophet for the same reason as the false prophets did it long before Jesus. A false prophet is loved by people appreciated and people will buy their lies. It's easier that way than to meditate and dig for truth by yourself.

Took me years to unlearn what I knew and what I was told in schools. I use to defend evolution with all my heart because I believed is correct. The more I learned and questioned, all the pillars of evolution fell like branches in the wind. Was nothing left to lean on or hang on to.

But I should not ask for attention and my own search for truth. All I wanted to do is to share what I found out after years of meditations and research and reading.

Day after day there were new questions coming to my mind. Many answers could be used to answer. Not one answer would please me or make me feel that "finally this is the correct answer".

I was very close to give up even if I was sure I was on the right path. Was too much, too difficult, too complicated, too this or that.

Time went by and I tried to accept "what will be, will be" strategy.

Standing there wasn't easy either. The need to know would make me unhappy at all times.

I would listen to all TV preachers and wonder how can they be so convincing while was clear to me they did not finished reading the bible yet?

Watching scientific TV shows or reading scientific articles of the last moment did help in a way. The most aggravating aspect to me was that some would assume things and than interpret them as scientific. That's the worse kind of science.

TV shows about the way (s) evolution took place would have at every turn an assumption or a missing "link". This seemed to me worse than a religion to present as scientific such "assumed science" that claimed the assumptions to be scientific facts. More often than not, arguments of the same kind as Thomas Morgan had about mutations would culminate in an explosive impossible to prove scientific idea.

In moments like that was easy to envision a time when modern science will have to apologize to the great public out there. At the same time science will have to go through a great reform.

The other day my "friend" told me that I need to educate myself and I should read "The Selfish gene" by R. Dawkins.

Did sound like an interesting title. While reading it I was amazed to see how much speculation was in that book. In fact the author himself would admit that what he says never really takes place in nature. Than "The God delusion" by the same author was recommended to me. I thought, finally a smart educated PhD who knows what is talking about. That book, by being admired as scientific by most people of science, could only mislead them.

Again the same pattern as before. Have to admit I did not finish reading it because made no sense at all. If it did make any sense was that more often than not the author would admit to some kind of design or preset pattern.

That aspect wasn't analyzed at any point in the book. I did wonder why would anybody admit that looks like a design, but continue to assume it wasn't? Or is Dawkins playing a double game? He wants to show that the design is present but tries to hide the designer? Maybe he tries to undermine the science by talking about such design? I am just being ironic at this point.

Again, all I saw was the acute need for modern science to apologize for trying to mislead so many people.

If Dawkins' books will make it into the scientific world maybe there will be few honest "people of science" to ask the critical questions. If not, once again, a doctrine will be accepted by the world of science. This one might last longer than Darwin's because doesn't give any kind of direction in particular. This doctrine doesn't need auxiliary theories to be supported. In this case people of science could afford to assume anything they want to and still be considered scientific.

Somehow Dawkins might become for scientific world what Paul was for the religious world.

Without Paul's letters there would be no New Testament.

Dawkins could give the New Assumption theory to the world of science.

Nor Paul nor Dawkins really cared to what extent they will mislead others. Both tend to defend the wrong side of their "modern" ideology.

Paul knew what Jesus did was possible, but because he wanted to take peoples' minds off the resurrection of the body idea, he served them some misleading stories. Would talk forever about himself, how he did this and did that to the Christians, but was never honest enough to explain exactly *why* he did it in the first place. After all he knew Jesus while he was alive. Some say that is not true. But is it possible that Paul did not hear about Jesus resurrection and the big decision Caifa made? In fact he might be the one who really kept after Caifa to kill Jesus. Well, that's an assumption based on the fact that Paul tried to disorient Peter few years after Jesus went thru the great transformation from physical body to spiritual body. If one, reads carefully his letters is rather clear that Saul/Paul did not like Jesus at all. He tried for years to kill all his followers. He was able to travel long distances only to get rid of them. When he did realize that's impossible in his generation, he did the damage in his letters for future generations.

Dawkins on the other hand he does "sense" there is a design but tries to give room to plenty of assumptions in such a way that evolution could still be "alive" in those assumptions.

For someone who claims he has so much education, and knows about genetics claiming this kind of contradiction is rather obsolete science.

For sure he talks very little about genes in his "The Selfish Gene". He did not explain in the "layman's" view the critical aspects of cellular division process, which is the only way the selfish gene could be "selected" for.

His ideas might have a large audience in the world of science, because very few care to understand these aspects. Even if these cellular division aspects would be understood they have to be followed by a complex and systematic work in order to get a clear picture about it.

Nobody will do that work and be able to explain it to the majority that doesn't care. It is way easier to embrace a misleading theory, especially since is so confusing that even the author admits that the kind of models he presents can't be found anywhere in nature.

I wonder if I should make this book a long detailed book or just to give a succinct idea of the basic relation between the fields of study. Yes I do consider The Holy Bible a field of study. After I've read the Bible few times, was easy to conclude that, is the most scientific book humanity has so far. Its "fiction" will be our tomorrow's science. That's exactly what Paul tried to waste. He had a lot of knowledge but wasn't able to get to it as fast as Jesus did.

I am very much inclined to stay with the short story because "my sensors" feel the futility of my effort. My "credentials" are not as impressive as Dawkins's are. My work of science didn't even take me to the point to publish it. I was told that is too much against the "main stream".

If one doesn't know how to go with the flow will be crushed.

Again this book is against all the main streams. I do not use any ideology in order to get the attention of any particular group or their support. Searching for truth, as sarcastic as may sound, is far more important, since the only "crossroads" I can go to are the books I am allowed to.

To try to stand-alone and separate oneself from any main stream is suicidal indeed. But for those who care to know more that's the only way.

So the book is mainly for those who want to search for themselves.

These minds need very little guidance and will not embrace the already shaped up ideology. The majority will continue to stay with the main stream doctrines, be that the religious or the scientific doctrine.

Very few will admit that science is a doctrine. To see science as doctrine I do recommend reading Dawkins's books. His books are but doctrine. He is far from modern science and even from what Thomas Morgan knew.

His books give more room to assumption than any other I read so far.

Religious doctrine does change over time in the sense that the church won't lose too many people. As soon as a religion speaks truth there is nobody to follow that religion. Islam, the new Islam religion, accepts science to a certain degree, which is why some would rather go with

Islam. This religion is more flexible from scientific point of view, but is more corrupt from spiritual point of view.

Mecca the place where Mohamed was born, and expelled him and had wars against him, is the place where Muslims love to go to worship. Do they know why Mecca wanted to kill Mohamed? Do they care about the fact that Mecca had idols that were worshiped long before Mohamed?

While Mohamed said; 'no idols should enter in your religion', Mecca was the heart of idolatry in his time.

Mecca should be the place Islam should go there to destroy the memory of idols not to worship. How come so many educated people who are part of Islam do not point out this aspect about Islam? The moon was one of the symbols of idolatry in Mecca.

Should one care to know what they worship?

Every prayer time is towards Mecca, Mecca that tried to kill Mohamed. Nobody seems to care what he or she worships, for as long as the "main stream" covers them. Should one obey a religion that doesn't care about what it's only prophet said or went through?

Is such religion pleasant to God/Allah?

This is why nobody can touch the main streams, because nobody cares to know what lead to the main stream.

Let me depart from the main streams, because are way too many, and nobody cares about their roots and the improper influences they suffered over time.

A 500 year ago Christian would not recognize his own religion today.

If Mohamed would look at what happen to his teachings and his book in more than 1300 years would not think is one and the same.

Maybe that's why some books were not included in the Bible, so the main stream religion won't have the chance to interpret anything in these books. Were excluded for a reason from the Bible, but that would benefit our generation by not having too much interpretation on their content. So when we will find and read carefully those books would be easier to understand what the author was trying to deliver because is not already in the frame of some dogma.

If The Book of Enoch were to be accepted in the Bible in 364 A.D. at the "Council of Loadicea", maybe to this day the Bible would be read only in Latin. Nobody would know to this day what the book really says.

Religious interpretations would be the only choice for those who can't read the ancient Latin.

The church did a good job by being restrictive to read The Bible only in Latin for centuries.

Why would I read a book to anybody if they couldn't understand what I read? Why would I interpret the text the way I please to them if is impossible for them to achieve their own opinion?

Could it be in such a case one is trying to corrupt the understanding that is suppose to come with that book?

Why was in fact read at all?

Simple to answer I think: Was read because people knew about Jesus in those times and His true message. The Bible couldn't be changed by much because the books that were put together in the Bible pre-existed the Bible. Those that voted for the format of the Bible dictated which one of the books of the prophets in the Old Testament or the pre-existing Hebrew Bible will get to be part of the Bible.

They had to use at least few of the Gospels about Jesus, because the event was still too fresh, about two or three generations after Jesus.

The only option left, was to use the books as they were, and preach them to people in a language they would *Not* understand. Of course, in 364 A.D. the Roman Empire was a good excuse to do so. That habit was still enforced centuries after the Roman Empire fell apart. To this day some churches will read the text in Latin too.

Why is it that many were killed and the books destroyed if anybody dared to read them outside the control of the church?

Mendel read forbidden books and that was the reason why he was expelled from his monastery. At least this is the claim in some versions of his biography. He found out things nobody was allowed to know about. Not only did he find out about these books, but also he managed to read one of them. That's why his experiment did not need the modern science's knowledge of his time. You'd think by now that I dare to assume all this. But I have plenty of proof and I know there is more I just can't touch it yet.

For instance; Bateson claims that; "his experiments are worthy to rank with those which laid the foundation of the atomic laws of chemistry". Bateson is talking about Mendel. Others compare his discovery—"of an importance little inferior to those of a Newton or a Dalton."

Mendeleyev is another one of these great pioneers. His knowledge about chemistry had no precedent. He made a sudden turn in chemistry. Without his discovery the laws of atomic chemistry would not be possible. As strange as it sounds not even Mendeleyev himself knew for sure what was it, he did discover. That's why he wasn't able to complete the table.

Newton made a sudden turn in physics and mathematics. Nobody knew before Newton what he "discovered". The opposition in the world of science of his time was so drastic that he did not dare to share everything he did "discover". His main frustration was in the end with alchemy.

If Newton had Mendeleyev's information with no doubt he would've been able to put together real chemical reactions that could lead to elemental chemical change. To this day that transition, which is called "alchemy", is not "discovered" yet. Some claim they know alchemy but our modern science doesn't agree with their explanation, which is too metaphysical for modern science.

Newton's immortality "stone" apparently wasn't perfected.

Many are still looking into it. Very few will claim it as main stream modern science. Their chances are very limited to find it since one main ingredient is missing from their knowledge. The missing "link" is the part Paul knew about, when he said he would tell Corinthians a "holy secret". But he told them only its principles not the actual phenomenon. Was that secret part of the Holy Grail?

Nor back then nor in our days anybody could claim they understood what Paul said in Corinthians about the transformation of the body.

Again I was prepared to go into each of these "great minds" biography and bring to you more information of how they managed to reach to such conclusions. I find it to be a waste of time since there are few biographies on each of them.

In order to search which biography is more accurate would take me a great deal of time and just get lost in nonessential details.

If I would have enough money I would pay someone to really dig in the moment when the "discovery" took place.

But again, is it really my duty to find the truth about these basic aspects or is it the duty of our modern science to go ahead and do its duty in searching the truth?

In Newton's case the work was almost done, since was clear at some point he did claim that he bought a set of ancient books. In another sentence he also claims that he "stands on the shoulders of the giants". To this day many see the "giants" he talks about as the ones before him in the same field of study. As we well know by now, nobody before him had such information. Or if they did have it they did not manage to present it.

Could it be that the giants he is talking about are the same giants Enoch talks about?

Many times I had doubts about these giants. Many times I tried to explain them in the simplest ways.

But I failed to understand who they really were.

The aspect that bothers me the most was Newton's fear of church. His refusal to believe what he was told by his own authority and he became very tacit about it. In fact he hides from them what he did discover. His students thought he was mad. According to some, his students did not care to know what he was teaching. His knowledge wasn't exactly main stream knowledge.

Then and now most students are in school only to get their diplomas. That aspect about education is true to this day.

If Newton didn't come to be part of the royal society, maybe to this day we would not know about his discoveries.

There are no mistakes. Wasn't a mistake that he got those ancient books. Wasn't a mistake that he did not get the same books Mendel did. Wasn't a mistake that Mendeleyev has the information for the elements our world is made of, before anybody else did. The idea of atom is ancient. In fact is mentioned in the Qu'ran in few places.

If the ancient Islam (before Mohamed) knew about atoms, should I assume that Paul didn't?

Or was Jesus without knowledge about this kind of matter? But if they knew it, how could they teach others all that knowledge?

In that regard maybe when Jesus says "You'll be better than me" he means you'll know more than I do.

Our modern science, with no doubt, has no way of stopping. Will meander for a while, but will find its flow soon.

In order to understand our modern science better, we need to reconsider the roots of our modern science.

What few had the chance to "discover" about main phenomenon on this planet already is main stream knowledge.

Those discoveries were no accidents. Not all of them claim, as Newton does, that they found their knowledge in ancient books.

Mendel didn't claim that out of fear or pride. He just tested what he read. The book told him "it will" segregate.

It did segregate. He could not understand why it did. We can't understand why it does, but it does for every character. Well not if is not a hybrid, as in the infamous case he did experienced with one of his plants. He did not know why it did not segregate at that time. In our times we know why it didn't.

Oh, I think I omitted to tell you that part of the story. Yeah I tried to get away with it, but you got me since you know Mendel's experiments and how he was treated by his time "modern scientists".

Maybe I should try to explain to a degree what it is, Mendel really "discovered".

I will not use exactly his "experiment" settings so those who think they know what this is all about, will just skip this part.

Let's say we have a mongrel. What's a mongrel? A hybrid or mongrel is a genetic combination of two different characters for the same allele. Red eyes and blue eyes is the same character, which is the color of the eyes. If "red eyes" is a recessive character will manifest only when the two alleles are recessive for that character. Like albino in black skin people. If the mother and the father have albino recessive, there is a good chance that couple might have an albino child. So the hybrid/mongrel for F1 (first generation) has red eyes recessive and blue eyes dominant. All the fruit flies will have blue eyes in F1.

The segregation begins in the F2.

If we use this blue eye fruit fly, that in fact has the red eye recessive in F2; the segregation will be of 3:1. In other words if there were 20 fruit flies, 5 will have red eyes and 15 will have blue eyes in F2. If the 5 with red eyes will have many generations after F2 all will be red eyes. If the 15 that had blue eyes in the second generation will have generations after F2 the segregation will continue in different ratios. There will be many combinations of red and blue eyes, but about ¼ will go back to blue eyes dominant allele. The 50% will continue to segregate for many generations. There are different results for different characters of how this segregation will continue that's why if you care to know more about it you'll have to do your own work. Genetics 101 is a big book and about ¾ of it is only about Mendel's laws.

All segregation means, is that the WT or original genotype, will tend to regain its initial "purity", which in fact could lead to self destructive action. Consagnvinization (same blood) is known to lead to self-destruction.

These aspects are still not fully understood by our modern science. Every research that was done on these aspects ends up with uncertain results.

For some weird reason modern genetics prefers to use all kind of genes transfers from one specie to another.

What modern science does is actually happening in nature at all times. All we need to do is to select for that specific character. Rather difficult to explain how that selection goes, but is not the kind of selection we know about from our biology class.

Selecting for red eyes from the 50% combo, doesn't mean that we select for some character did not exist before and could lead to new specie. Simply means that in the 50% combo over time we could get the double recessive genotype for red eyes. By having a red eye fruit fly from a blue eye population won't be by any means a new specie of fruit flies.

At any rate, would be too much to explain that sentence to someone who got A-s for learning different.

The law of segregation in itself shows that the character had to be present before could be selected for it. The character is part of that very specie and after selection will still be part of that very specie.

The fact that one gene can be transferred from a specie to another does not mean the genotype was altered to the point to obtain a new specie.

Is more like moving a brick from one house to another house. If the brick is slightly different won't fit in the structure of another house. In order to fit has to be cut the appropriate space for it. Even so the house is still the same, only got a larger blue brick in it.

This new gene/blue brick might be lost in the second generation/s or could render that particular individual unable to breed.

The point to be made is that; due to segregation, there is a moment when the hybrid will not maintain what was gained due to hybridization. So the hybridization is not an "improvement" to the variety, if it was an improvement, than indeed could lead to new species.

Again these aspects are not the main focus of our modern science; therefore we will never know how they really work in "real life". All we have as modern science are a lot of assumptions, just like Dawkins has, that don't fit the pre-existing laws.

Modern genetics plays with genes trying to cut and paste genes from a specie to another. This technology enables us to change a character at a rapid pace compared to the "primitive" selection methods. In many cases the same gene was found naturally in different species.

A modern science case came to mind. I think was in Time magazine few years back. The researchers claimed that they created some genotypes, which proved to be a false statement. Many took a big chunk of money out of that lie. Finally someone had the guts to claim that was impossible to obtain what this team claim they did, so the case was or still is in trial. The main researcher admitted to some kind of "scientific fraud". Hmm now I sure have to look it up since I brought it up. If I forget I think it was in 2003 or 2004 January number.

I wonder what would've Darwin say, if he knew about the law of segregation.

Another claim I just saw on the news is that "new life" was created in the lab. Nevertheless the announcement is not "official" yet. The newly created life in the lab needs a type of bacteria from real life to survive. That bacteria better be alive when the new life will be transferred into it. I just can't resist not being ironic about so much arrogance in the scientific world.

But let me try to move along, since the purpose of this book is not to figure out what science has to be focused on.

Or is it?

This is what Mendel said on his deathbed in 1883—" *My scientific studies have afforded me great gratification; and I am* **convinced** *that it will not be long before the whole world acknowledges the results of my work* "

Sounds selfish, but being convinced that his results will be part of the future science is not characteristic to scientists that were rejected for more than 40 years by the main stream science. To this day our science can't understand these laws of segregation. Yet, Mendel was able to discover the laws and be *convinced,* while his work was rejected as scientific in his life time.

This is another very intriguing aspect of Mendel's type of science.

To be convinced 40 years before the science will accept your work as scientific. Yet the science is not able to understand what triggers these laws, since they are totally counter intuitive if looked at from evolution's point of view.

What made Mendel to be so sure about his laws?

Could it be he had parts of those "ancient books" Newton is talking about?

One aspect from the Bible I would like to cite here is Genesis 30-25 to Gen 31-13.

Who cares to read it will see the whole set up for a very interesting genetic "experiment" that has known outcome, otherwise wouldn't be set up at all.

I will actually cite only few verses; *Gen 30-34-41 ;" The same day he removed all the male goats that were streaked or spotted, and all the speckled or spotted female goats (all that had white on them) and all the dark colored lambs, and he placed them in the care of his sons. Then he put a three day journey between himself and Jacob while Jacob continued to tend the rest of Laban's flocks.*

37 Jacob, however, took fresh-cut branches from poplar, almond and plane trees and made white stripes on them by peeling the bark and exposing the white inner wood of the branches. Then he placed the peeled branches in all the watering troughs, so that they would be directly in front of the flocks when they came to drink. When the flocks were in heat and came to drink, 39 they matted in front of the branches. And they bore young that were streaked or speckled or spotted. 40 Jacob set apart the young of the flock by themselves, but made the rest face the streaked and dark colored animals that belong to Laban. Thus he made separate flocks for himself and did not put them with Laban's animals. 41 Whenever the stronger females were in heat, Jacob would place the branches in the troughs in front of the animals so they would mate near the branches. 42 but if the animals were weak, he would not place them there. So the weak animals went to Laban and the strong ones to Jacob."

Doesn't sound to me that Jacob did a lot of trial and error experiments in order to get this kind of knowledge and be so sure of being correct.

How did Jacob know about all this somatic mutations and was so confident, so convinced, so sure of himself that did risk the well being of his family on such "experiment"? He was the one asking for such wages from Laban, since he knew that Laban will lie to him about wages. Yet, Jacob was almost as certain of himself as Mendel was. As irony has it, both dealt with phenotypic characters, colors or shapes.

The difference between Jacob and Mendel is that he does explain how he knew about it.

Mendel claims that was his own studies. Yet, Mendel, just like Jacob did not run the classic frame of an experiment. Did not have to find out what is it they want to study on. Both knew with certainty what to expect before they started.

That's why Mendel even though had a huge failure in one of his experiments and he could not understand why that did happen, his confidence was as high as can be on his deathbed.

One of Mendel's experiments was run on Hieracium or hawkweed. This plant from Asters family, has asexual reproduction, therefore can't lead to hybrids or mongrels. The law of segregation is valid only for hybrids.

Now let's go back to Jacob for a moment.

He banks everything on his ability to induce a somatic mutation. Which, by the way, he had no clue what that really is in genetic terms. He only knew the goats could get these spots or streaks on them.

When I read this part I did find that was a fascinating "experiment". Was thinking for a long time at the details he gave. Some of them did not make sense to me. But some would match scientific discoveries that were made only few decades ago. Our modern science found out about the possibility of induced somatic mutations about 40 or 50 years ago.

Jacob was the 23rd generation from Adam and did perform induced somatic mutation.

To this day, the best way to induce somatic mutations is the use of some plant extracts, besides radiation and temperature or pH shock.

Jacob was told how to use plant extracts in drinking water for animals to induce somatic mutations in the first generation. Removes the bark and adds to water the secretion/sap of the fresh wood in a certain mixture.

Unlike Mendel, Jacob tells his wives how he found out about it. Well, Mendel was a monk, so he had no such option. What Jacob told them is recorded in the Bible in;

Gen 31-10-12 "In breeding season I once had a dream in which I looked up and saw that the male goats mating with the flock were streaked, speckled or spotted. 11 The angel of God said to me in the dream "Jacob" I answered, "here I am" 12 And he said, "Look up and see that all the male goats mating with the flock are streaked, speckled or spotted, for I have seen ..."

Seems very unclear what the angel of God really showed Jacob. This could've been true only in part. He told only part of what happen in that dream so he can convince his wives to follow him. By telling them that an angel gave him directions on how to induce the somatic mutations, his wives would trust him and follow him. That could be one's opinion. For some reason I think Jacob tells the truth.

It's easy to understand why Mendel won't claim that what he knew was shown to him or he read about it in the "forbidden books".

If a scientist would run such an experiment without any previous information would give up on it right in F1 and conclude that hybridization is impossible, since all the F1 individuals would have the same phenotype. The dominant allele would be the only one that shows to the eye of the researcher. The first reaction would be;—the experiment was a failure at some point for some reason-.

Mendel was very aware of the F2 segregation. He kept track of his seeds with very accurate numbers. That in itself is rather impressive. Is also impressive the fact that the world of science didn't care to accept these obvious and inexplicable laws for more than 60 years after publication.

These two aspects are very intriguing to me. Mendel has to get to the F2 in order to see results, but Jacob could induce mutations in the first generation. That's a big question for me. Is it possible that by chemical activation of some sort Mendel could've induce the green peas color to yellow? If that is possible what does that really tell us about allele and the possibility to mutate? Should science look into this aspect more than is looking into evolutionary aspects?

To this day nobody can explain or understand these laws, but at least in our times are accepted. Many would claim that Mendel had time to count peas that's why he was able to find out. That would be strange to claim, since his discoveries weren't considered science at all in his time. Mendel knew that what he discovered was correct and scientific, but took decades to accept these obvious laws so definitely was more than having time to count peas.

These laws are the basis of our modern genetics. In fact I just revisited a book of Genetics, more than half the book is only about how these laws are taking place at different levels and characters, but there is no word why this segregation takes place with such precision.

Yet, Mendel claims these laws are his own studies. Was it out of fear or pride the reason why Mendel did not tell the truth about his unusual discoveries? I know I repeat these questions, but are worth repeating them.

This was a brief parallel between the first "true" recorded geneticians on this planet, Jacob and Mendel

At the same time is known that in ancient Egypt the magic practicing priests knew about somatic mutations and healing by means of herbs or plant extracts.

These kind of scientific discoveries are all over the map for our modern science. Many famous names could be linked to these "ancient" sources of information. In order to do so I would need to spend a lot of time to read between the lines in every biography.

Nostradamus admits that he used ancient books to predict the future. He had to destroy those books since the church was after him when they found out he had some of "those books".

If read carefully, each biography of the pioneers in certain fields of science, there is a certain enigma that can't be explain in what they say or what they think. Just like Mendel's deathbed affirmation. One doesn't expect to hear such affirmation after he was rather excluded from science and even proved wrong with his "asexually reproductive" hawkweed. That experiment was considered "fatal" to Mendel's scientific career. Yet, his confidence wasn't shuttered by such "minor" fatality. Of course he could not understand how that happened, because the asexual reproduction was discovered by accident later on.

Imagine if the then "modern science" would consider analyzing Mendel's results carefully what would've happen to science itself. They would shift the entire scientific thinking right then. The focus of then science would've been around Mendel's laws and the factors that determine this phenomenon of segregation. Evolution would've been forgotten while in its crib. But evolution was a more hip theory than the cause of segregation.

Today we would have a completely different approach to biology and most life sciences. But the law of segregation was considered an error and science continued to be stuck in the "main stream" science to this day. That's why to this day no field of science can explain why the segregation takes place in the same ratio for any given character. This "why", could be called the "critical why" of modern science. Once the "critical why" has an answer that will mean the "fall" of scientific doctrine.

I would like to analyze Newton's work, but for the purpose of this book he made it easy for me. He admitted at some point buying a set of ancient books. Those books were his secret and made him the pioneer he was. He standing on the shoulders of the giants meant the real giants, not the ones who hacked his work or the ones who saw the universe as a mechanical system like Decartes.

If we accept what Newton says about his own source of knowledge there is no argument of how he "discovered" the laws of gravity.

To this day modern science can *Not* explain what gravity is. When Newton was asked what is gravity he was veeery angry that he couldn't answer that question.

Modern science doesn't know what gravity is but knows its laws from Newton's discoveries.

Modern science doesn't know why segregation takes place with such an accurate ratio for every character but finally accepts that indeed segregation does obey that ratio.

There are many examples of this nature, which are the true bases of our modern science, but if one understands one or two, already understood a hundred or more. If one rejects one example, a hundred or more will still not be sufficient to understand this matter.

The mind set in our modern science at this moment in time is still as it was in Mendel's time. All his experiments, all his results meant nothing because didn't fit in the mentality of those who read them. All his results didn't fit in the then and now theory of evolution, therefore were what we like to call "counter intuitive" for his/our generation.

Few, who did study this effect of segregation, did point out that this effect goes against the very core of evolution, but their opinion never became main stream "science". Too many research projects going on are based on funds that have to serve the evolution theory in some form or shape.

Modern science still chooses to ignore basic, obvious observations like the laws of segregation, and prefers to look for twisted ideas about how evolution took place.

Unless one really cares to look at these aspects by him/herself and analyze them in a logic way there will be no change in the main stream "knowledge".

How could that be done?

The only way we have right now is very difficult. It's much easier to accept what you are being told by the main stream system of believes, than to analyze by yourself.

If one chooses not to believe anything at all, that could lead to total devastation. Of course, many people live for their idols. Or many people live to gossip, to shop, to bingo and much of the same. That could also be a system of believes. People love to get addicted to something, that something becomes the priority in one's life. Unfortunately these kinds of priorities will always lead to some sort of self-destructive behavior. This type of behavior is being analyzed by modern medicine, but so far only "special" places can help one recover from addictions. The focus of one life is shifted in these places. In fact people have to use these facilities if are victims of addictions. What is rather of concern is the fact that these addictions are formed by the social system we live in. These are known, obvious facts but there is no real main stream education for the new generation to prevent such addictions.

If you managed to read this book this far you are already an exception.

Your effort will be rewarded by some ideas that could be totally new to you.

So far I tried only to prepare the mind and soul for what you are about to find out.

I am not very good to prepare anybody for such a shift of thinking.

The worst part is that I can only show you so much and if you care you'll have to walk your own walk to find out more.

Let's stay for a bit longer with the events of our times.

Immense ideological movements taking place as we speak.

Christian world falling apart the way it was known to our parents. Islamic world is in the process of trying to enforce a religion of terror on the world.

Christians tried to do the same long time ago. In our times will be more difficult to do it for such a long period of time since news travel faster then ever before.

By having the news at one's fingertips people should be able to read and think about what they learn. Of course many choose to be sheltered from all the news or information. That could hurt them or could be a very good thing, depending of how the events will shape up.

If the war in Iraq will lead to more wars in the region is good to stay sheltered because the events are not stoppable at this time.

If the war in Iraq will stop or even come to stand still for a while that could help those who try to help others to understand that violence is not the answer to anything. Otherwise the violence could be the only answer left for humanity.

God loves to punish us with our own choice. The entire Bible is about such events that took place on this planet.

God send messengers asking people to stop violence to stop hurting each other. People can't stop once they start the cycle of violence.

Every time God destroyed a civilization was due to their never-ending violent actions.

Obeying the golden rule is the only main requirement God has of us. Since day one, humans could not obey that rule. Everybody would love to be treated with dignity, but won't make the effort to do the same in return, rather choose to kill those who won't do it. Respect me or else, kind of mentality.

This has a simple conclusion; Humans are not able to lead themselves towards their own desires.

How easy would be to choose an education system that doesn't lead to humiliation of others? If the education system would get rid of the old clergy like format, those who choose to learn would feel the freedom and joy of learning. The educational system is designed in such a way to kill that feeling at an early age. Once that natural instinct of knowing more is destroyed, learning becomes a painfully frustrating and humiliating experience.

Cheating will be a form of surviving in such a tense atmosphere. Desire to get out of schools would become priority in one's life. Later years will bring that natural instinct back, but by then for most will be difficult to reengage in the education system.

Could a new education system lead us to the truth each of us would like to know?

Each of us at some point in our lives would like to know the meaning of our life and once understood, to make the decision to walk in or toward that meaning.

I use to deny that desire for truth.

What type of education could give us the option to know about ourselves as much as a human mind was designed to understand?

The simple, non-competitive type of education could be the only option we have. The kind of competitive education we have leads to frustration and systematic rejection of knowledge. The competitive education is the best way to keep humans away from knowledge and kill that natural instinct of knowing more. At some point or another for some reason each of us will experience frustration due to competition. Once that feeling gets to us there is no escape. One will give up or choose to cheat or choose to fight in any possible way, but not by finding pleasure in what one has to study.

On the other hand, those who realize how wrong and confusing education could be, might choose to step aside even if they have the ability to achieve a lot more.

While I was in school there were many instances when the professor could not explain to the class' expectations the topic which we as students were expected to know.

The grading was just a joke. Even the professor knew that what he was teaching us was not accurate. Yet, the grading had to go on.

As outrageous as it sounds, grading is the main reason why truth can't be thought in schools. If a student would finish college and had no grades at all to show by the end of 4 or 5 years of college there would be no criteria to asses his/her ability.

A careful analysis of the grading system and its flaws might show that grades will never reflect one's ability in any given field of study.

One sick day could ruin one's grade in one or more disciplines. "Bad luck" is also critical when one gets graded. Grading makes professors, gods over students. Meanwhile the student should dominate the professor with his questions and desire to know. How many students would dare to debate with their professor and still hope to get a good grade?

Every school and every professor will have a different method of grading. If one finishes a school with a "flexible" grading system could have a score way higher than the worst student in a severe grading system. In real life the one that had the worst grades in a rigid grading system, might perform better than the one who had better grades in a flexible one.

When the church came up for the first time with this system of grading was probably intended to control the students. At that time the amount of information was rather minimal. Reading and writing was the main core of education. In fact that was the main reason why the church did allow education to teach people to read and write the religious obligations they had.

If one knew how to read and write that opened the door to long distance communication. The church would send for money and the folks out there would know how to read what the church desires of them.

Would make a lot of sense to educate the masses and make the job of collectors easier and more control over people would be made possible.

The limits were imposed by church. The books of science were forbidden. Those who read them had to hide that or they could risk their life.

The library in Alexandria was a place that had a lot of science, therefore had to be destroyed in order to keep such information away from masses.

Many books that were spread allover the place became the targets of inquisition. Even the Bible at some point became *too scientific* for majority out there, according to the religious leaders. Who dared to read the Bible by himself or in groups outside the church were considered heretics and had to be killed and the books be burn. The church made no secret of such actions.

If one chose to read more than was allowed to, would be killed in a public place. This kind of action would scare to death most people, so they would not care to read more than they were allowed to know.

This could be another reason why Jesus saw as futile to write anything. He probably knew that if gets killed what he writes will be destroyed too. At the same time, in my opinion, Jesus did not write for a different reason. I'll talk later about why Jesus did not write.

Humanity will always have to fight for freedom of information and freedom to think "outside" the imposed box.

If the imposed limits had been defeated back in the times of Jesus, humanity would be better off by now. At the same time knowledge could lead to self-destructive actions as well.

As soon as science comes up with some out of ordinary discovery, the government will use that for destructive purpose first. Germany and the World Wars are a good proof to this type of human behavior.

Just like little kids when they discover fire and its potent fascination. Playing with knowledge could be a lot worse than playing with fire.

As you read these lines our world is in the self-destructive process. Missiles, is a "modern" term again.

The effort to bring peace to the world is just the job of few devoted militants. They will never get billions of dollars for their cause, but they might get jail time.

We shouldn't wonder if God would destroy our educated world. Education served only our destructive side.

Blaming science for all the human evil action is not an accurate prospective either. Religious world blames knowledge for all humans' evil actions. Scientific world claims the opposite.

Science is just a double edge sword.

The day religious world and scientific world will come to understand the roots of science and why was it oppressed for so long, might be near. If the world will choose a doctrine of science and the type of confusion some scientists try to mislead to, humanity will see for centuries to come the same "history repeats itself" pattern. By being so advanced, the scientific world could only go towards destruction if continues to stay in the scientific dogma. I know, these kinds of affirmations might not make sense to many, but is not my fault that we were educated to believe otherwise.

This time frame it's a good conjunction of events to re-direct ourselves.

The religious corruption is public knowledge. The scientific doctrine is floating like a two by four in murky waters. If science will choose to "attack" now and become dominant over any other ideology with no doubt the destruction is an imminent event.

The other alternative is to search for the roots of science. Once we find out the truth about it, there might be a good chance for harmony on this planet.

Somehow Mendeleyev's mother words come to mind. She wanted him badly to be educated and accept the divine intervention in their life. That kind of mentality is out of ordinary even in ours times. Well, actually mostly in our times, since people of science have the certainty that they know everything there is to know. Most scientists don't see the room of divine intervention in their scientific life. Mendeleyev himself did develop a "non scientific" philosophy that lead to be self expelled from the world of science. That was the time frame he did discover the most critical scientific ideas. His "periodic" table came as a "revelation" to him. At the same time he did "predict" that the missing elements do exist. He predicted few of them with great accuracy and others in within a very small limit of error. That limit of error is still not sure was it his error or our modern science made the error.

Yet, he was not accepted by the world of science in his times only after a while. He wasn't, according to most around him, a brilliant mind. He was born in a remote Siberian village. Was part of a large family. His mom was on a "mission" with him.

Her "philosophy" won't fit either the religious ideology or the scientific one of her times. She could not imagine humanity making any progress without God's help and guidance. Her son discovered the most critical aspect of our modern chemistry. Students still use the periodic table of elements, even if nobody knows exactly how that table was discovered.

One could write entire libraries about how out of ordinary these "great minds" really were. Yet, to their own generation, they seemed like ordinary people, maybe even below average.

So far I made few allusions about the roots of science.

The ancient books Newton bought and used.

The books Nostradamus had to destroy in order to avoid punishment.

The Library in Alexandria was burn to the ground by Christian leaders.

Mendel's convincing statement regarding the results he got.

Mendeleyev's out of common discovery regarding the periodicity of chemical elements.

All these discoveries and many more that weren't mentioned should trigger the attention of the inquiring minds of our times.

In the longest time our generation has some freedom to think and reevaluate certain scientific aspects.

Should we loose such freedom over any given ideology?

Took centuries for women to gain the right to learn and express themselves. Still in its infancy, this freedom should be used carefully.

Took centuries for the ancient knowledge to resurface and be of some use to humanity. Only in 1800s the great discoveries came to light.

An inquiring mind should not just accept these aspects about our modern science "as obvious".

If one chooses to learn more will have to read all the forbidden sources. Analyze such aspects and try to see why it is impossible to fit them in the fabric of our ideological grail/network.

Why is it impossible to have religion and science under the same "roof"?

Why did kings need to use the information given only to prophets?

How come the great minds of science claim they had revelations just as Jacob claims about his somatic experiment?

Why would all of them become pioneers in the same time frame, which was around the mid 18th and the beginning of 19th century?

With no doubt, "the book of questions", could be written regarding these aspects. Ha, you mean to say this book has too many questions?

Let's look for some answers than.

One source that I find irresistible is The Book of Enoch.

Many do think that this book was written well into Christian era. The book was found, indeed, in different versions during Christian era. Most of its versions were destroyed. The Ethiopian and Slavonic versions did survive. Maybe the Slavonic version was hidden in Siberia. I am sure there are other versions that were not "found" yet.

These versions are based on the book Enoch and eventually the parts that Noah wrote about his great grandfather.

Enoch was the 7th generation from Adam and was Noah's great grandfather. The book is rather explicit about that.

Abraham lived ten generations from Noah. Noah was the 10th generation from Adam.

These aspects will just pin point the time frames. The genetic aspects of these generations were discussed in another book "Sci-Fi Bible".

If one reads carefully, the Book of Enoch will understand why this book did not make it to the Bible in 364 A.D.

This book according to Tertulian was in fact the "Bible" the generations before Abraham used it as the "Bible".

The Hebrew Bible somehow avoided this book as well. Meanwhile this was the Book Abraham read to find out about his God. As one recalls in Abraham times the Egyptian idols took control of the culture. Abraham's father himself used to make idols and sell them. Some might know the story about Abraham trying to explain to his father why his idols are not God.

I believe I told this story in another book, but it's a good time to repeat it.

One day, while his father was away from home, Abraham did enter his father shop. He destroyed all the statues in the shop, but one. The one left was the largest in his father's idol "collection". When his father came home and saw all the gods "killed" asked Abraham—"Who did this?" Abraham said to his father—"the big god did it".

Terah, Abraham's father answered "-he can't do such thing". Abraham asked him "-than how come you sell these gods to do things for people?"

Terah got his son's drift and asked him to leave the house immediately and he refused to learn what God Abraham was referring to.

Hope you are ready for the information the Book of Enoch has for us. If you read the book, as I said earlier, you might know why this is not part of the Bible or the Hebrew Bible or the Koran. Why this book was rather "lost" for generations won't be difficult to tell.

As I read the two versions I do not think there is a big difference between the two. The main ideas are present in both versions. I prefer to use the Ethiopian version in this book.

It would take too much work to cite all other potential sources, which in a way or another, might be corrupted over time. The Book of Enoch besides the fact that has some missing chapters is maybe the most unaltered source of information. At the same time is the most difficult to understand as well. The reason could be the fact that in fact Enoch wrote many books, 366 to be exact, according to his affirmation. So far only two versions were found. There could be more, but I do suspect that might be known under different names (ex. I Ching).

For instance, the way Enoch explains the gates of the sun and the way the day length changes in relation to the gates of the sun is very difficult to understand. Not even an astronomer would be able to follow carefully what Enoch says, yet he is very accurate of how the day comes to be equal with the night at certain times of the year. This was ancient knowledge about the sky and sun, but our modern science discovered the solar system and its revolving pattern only few centuries ago. Enoch writes about this pre-established pattern about 6000 years ago.

Many claim the Book of Enoch was written only when was found. This claim comes from the church clergy.

According to "scientific evidence" the book of Enoch was written in the 7 th generation from Adam which was about 5000 to 6000 years ago. More precise, according to me, was about 5300 years ago. That is based on the information found in Chronicles and eventually other chronological parts of the Bible.

If we keep in mind that the Book of Enoch was written about 5300 years ago, the information in that book is rather astonishing.

If you did read the Book of Enoch, from now on we will revisit some ideas in that book.

If you didn't read it yet, maybe the following quotes will be interesting enough to help you read it and understand why the church was never too fond of this book.

Well let me use first a quote from the Slavonic version, since the Ethiopian version refers only to "them".

This quote could explain who the giants Newton was talking about really were.

Since the two "versions" of The Book of Enoch are in a way different, I should use both versions.

In my opinion are not versions at all, but different books, but the experts call them versions. Enoch tends to repeat certain aspects at different time frames. Due to that fact the books seem to be versions.

But again, I am only an observer not an expert.

Slavonic Enoch Chapter 1

⁶*And there appeared to me two men, exceeding big, so that I never saw such on earth; their faces were shining like the sun, their eyes too were like a burning light, and from their lips was fire coming forth with clothing and singing of various kinds in appearance purple, their wings were brighter than gold, their hands whiter than snow.*

By "*exceeding big*" one could easily replace that expression with the dedicated word "giant".

Were these the giants Newton is talking about?

As I read the book, seems there were different kinds of giants.

Each kind would teach humans different things. This affirmation might seem out of place for right now, but later on might make a lot of sense.

To me what Enoch describes is rather looking as abduction by aliens. The only different aspect about it is that he saw giants, meanwhile our modern times aliens are rather smaller and darker looking than Enoch's description.

Plus who knows how many are honest about their abductions.

Let's follow for a while what Enoch did experience. The way he describes it is rather clear that he had no idea where he really was, but the two "exceeding big" ones were with him for a while and explained to him what he was seeing or experiencing.

In the Slavonic "version" Enoch writes very little about the other kind of giants. Are mentioned only as "Gregori" when the two (angels) with him explained to him when he did ask.

This is from Slavonic version Chapter 18:

[1] The men took me on to the fifth heaven and placed me, and there I saw many and countless soldiers, called Grigori, of human appearance, and their size was greater than that of great giants and their faces withered, and the silence of their mouths perpetual, and their was no service on the fifth heaven, and I said to the men who were with me:

[2] Wherefore are these very withered and their faces melancholy, and their mouths silent, and wherefore is there no service on this heaven?

[3] And they said to me: These are the Grigori, who with their prince Satanail rejected the Lord of light, and after them are those who are held in great darkness on the second heaven, and three of them went down on to earth from the Lord's throne, to the place Ermon, and broke through their vows on the shoulder of the hill Ermon and saw the daughters of men how

good they are, and took to themselves wives, and befouled the earth with their deeds, who in all times of their age made lawlessness and mixing, and giants are born and marvelous big men and great enmity."

So clearly Enoch sees these other kind of giants. Rather interesting is the fact that Enoch seems to be familiar with them from his life on earth. These are the giants that had children with the human flesh, yet, according to God they were of spiritual nature. They saw "the daughters of men" and had children with them. These are called in the Bible in Gen 6 the Nephilims according to some experts.

They were what Mendel would call the F1, the mongrels (hybrids). Their children should segregate in F2 (second generation) in the well known ratio of 3:1. According to other experts the Nephilims were the ones that could continue to reproduce and keep the spiritual and human nature.

The Ethiopian version has more on this so I'll give those details when I use that version. In fact to be honest I don't know how to continue and not mix the two versions. Hmm I wish I was a good writer, but you'll have to deal with my ability to put together this astonishing information regarding our human condition on this planet. While I write I see so much material for the true "modern science" that confuses my mind.

A branch of biology should study this interesting past of human nature. Oh, I forgot that science and God's creation don't mix in our modern era. In our times the grants are still used to find out the "missing links".

The Ethiopian version has a lot of details about these Grigori, and what is rather strange that doesn't call them Grigori. But at some point is very similar with what these Grigori did. In fact in the Ethiopian version are called Watchers. Their deeds are identical with those of Grigori. If I knew more about Slavonic I might be surprised to find out that the word grigori might mean watchers. I leave that to linguists, they need to earn a living too.

Seems will be easier to simply insert text from the Ethiopian version. There are notes as well, made by few experts like R. H. Charles as well as references from some Greek translation.

This will be a lot of work for linguists in all these translations for generations to come. If the 366 books will be found and translated and have them study by "modern science" we might finally get an understanding of what God was really trying to relate to us.

The following is from Chapter 7 and 8.

In the Slavonic version there was Satanail that met with his "employees" on Ermon. In the Ethiopian version gives a list of names of those who met on Mt Armon or Hermon. To me seems to be the same place.

"(6) The Aramaic texts preserve an earlier list of names of these Watchers: Semihazah; Artqoph; Ramtel; Kokabel; Ramel; Danieal; Zeqiel; Baraqel; Asael; Hermoni; Matarel; Ananel; Stawel; Samsiel; Sahriel; Tummiel; Turiel; Yomiel; Yhaddiel (Milik, p. 151).

10 Then they took wives, each choosing for himself; whom they began to approach, and with whom they cohabited; teaching them sorcery, incantations, and the dividing of roots and trees.
11 And the women conceiving brought forth giants, (7)

(7) The Greek texts vary considerably from the Ethiopic text here. One Greek manuscript adds to this section, "And they [the women] bore to them [the Watchers] three races–first, the great giants. The giants brought forth [some say "slew"] the Naphelim, and the Naphelim brought forth [or "slew"] the Elioud. And they existed,

increasing in power according to their greatness." See the account in the Book of Jubilees.

[12] *Whose stature was each three hundred cubits. These devoured all which the labor of men produced; until it became impossible to feed them;*
[13] *When they turned themselves against men, in order to devour them;*
[14] *And began to injure birds, beasts, reptiles, and fishes, to eat their flesh one after another,* [(8)] *and to drink their blood.*

(8) **Their flesh one after another.** Or, "one another's flesh." R.H. Charles notes that this phrase may refer to the destruction of one class of giants by another (Charles, p. 65).

[15] *Then the earth reproved the unrighteous.*
Chapter 8
[1] *Moreover Azazyel taught men to make swords, knives, shields, breast-plates, the fabrication of mirrors, and the workmanship of bracelets and ornaments, the use of paint, the beautifying of the eyebrows, the use of stones of every valuable and select kind, and all sorts of dyes, so that the world became altered.*
[2] *Impiety increased; fornication multiplied; and they transgressed and corrupted all their ways.*
[3] *Amazarak taught all the sorcerers, and dividers of roots:*
[4] *Armers taught the solution of sorcery;*
[5] *Barkayal taught the observers of the stars,* [(9)]

(9) **Observers of the stars.** Astrologers (Charles, p. 67).

[6] *Akibeel taught signs;*
[7] *Tamiel taught astronomy;*
[8] *And Asaradel taught the motion of the moon,*

[9]*And men, being destroyed, cried out; and their voice reached to heaven."*

This is not all the knowledge they did teach.

They wrote books for men in almost every field of science we know of. Their intention was to teach humans self-destruction. From abortion to weapons all they did teach was of destructive nature.

What seems rather curious to me is the fact that God used them as watchers. Satanail was known to oppose God's plan about humans. In fact, God explains to Enoch how He created everything and gives a rather detail information of how Satanail opposed His plan from the very beginning, because he did realize God wants to create another world/universe.

How much freedom does God really give to angels compared to humans that they did become jealous of human condition? This is just a thought about the former "world", before there were humans.

Is rather intriguing how easy it is for humans to choose the right over wrong, yet most everybody prefers the wrong. The outcome of that choice was since ever, self destruction. Any addiction is but self destructive. Yet, all the choices we tend to make are rather the ones that will lead us to addiction.

My mind always finds some sort of excuse not to work on this book. Like "this book will never pay off the effort and all the reading you've done". Or "is obvious nobody will want to read this book because every step of the way you blame or modern science or religions while you know that everybody is part of one of them". I have to say sounds very logic to me. These kinds of thoughts could take my wings away.

There are very good reasons why this book should be written. There are people looking for truth and suffering when they figure out that their religion doesn't say it all, or the scientific theories are kind of empty. This book could help many who are looking for more knowledge and that alone should motivate me. Is a major change to search for knowledge that nobody will give it to us yet, is imperative for our well being.

The other day I was talking to my daughter, saying: "..that nobody will care to read this book because is not the typical American thriller". She said: "I'll buy it". Her remark made me laugh since; of course, she doesn't have to buy it. But she had a rather serious face when she said that. For a moment that made me think that maybe she was really curious about what is said in the book. Lately she tries to ask me questions of spiritual nature that she never asked before. Maybe I write this book for her. Maybe I write it for my children since in the times they grew up I was an avid atheist. They had no chance to learn about God or go to church because of my own believes while they were growing up. Now I believe they were lucky in a way to escape at least one kind of dogma, so if they will care to read this book at least won't judge thru a religious filter. Again seems that I am looking for sympathy. Let me return to the previous ideas.

Earlier I made few remarks of how many books were destroyed by church leaders in the past. But that's no news at all. In Jeremiah's time the king destroyed right away what he wrote. He had to re-write it and try to run for his life.

It's in the nature of leaders to try to mislead the people. Does that sound familiar? I can't afford to give too many examples of this nature. My only hope is that this book could open few eyes and raise major questions.

All those books that were written by the watchers were never found by modern science. Yet, as I mentioned before, Newton claims he stands on the shoulders of giants. Also admits that he did use "ancient books" to discover what he did. Mendel seems was in the "vicinity" of some books that did explain the ratio of segregation of mongrels. The Greek text is almost clear on that ratio. But I am sure he used more than that text to come to a conclusion rejected by science for about 50-60 years. Meanwhile, Mendel dies confident that his source of information was correct, therefore his laws will be proven correct in the future. Newton also claimed that he doesn't really write for "this generation".

As I read the other day "The Sphinx and the Rainbow" by Loye he did mention that Carl Jung claimed using the Chinese ancient book of Changes or "I Ching" in his line of work. To this day he made great impact on psychology but is still not resolved. When I bought this book by Loye I was hoping to find a different approach, but with few exceptions was mostly a bundle of layers covering scientific discoveries and their failure. There was a remark that made me think again of how modern science operates. I should probably paraphrase since I did not ask for permission to cite.

'Few scientific discoveries were considered even in the recent years as "outrageous" since were against the main stream knowledge. Modern science had to reassess those discoveries. For now are seen only as possibilities.'

Here is only one example: in 1941 Dr. J. A. Stratton noted that "a particular field equation used in electromagnetic theory had two solutions".

The first solution was the so called "retarded potentials" or what is usually observed and expected.

The second "leads to an advanced time, implying that field can be observed before it was generated by the source". According to Loye; "heresy can quickly become orthodoxy" even in science, funny what a valid observation Loye made here. Due to the fact that science does embrace doctrine or orthodoxy, only by 1968 this "advanced potential" was considered by those working with electromagnetic fields. This was related to earlier ideas given by Paul Dirac. At the time Dirac was considered mad/crazy by his world of science.

Loye was right to say that science quickly becomes doctrine.

From 1931 to 1968, which is more than 35 years of our modern science, this basic idea was rejected just because its author has gone mad for a world of science that couldn't escape its own doctrine?

Maybe the doctrine should be seen in God's revelation to Enoch of how He created the existence from nonexistence or the visible from non-visible. After all this expresses exactly the same idea as Dirac claimed in his madness. Here it is what God told Enoch about creation, for those who are curious enough to have the whole picture.

This is from chapter 24 and 25 Slavonic version

"[3]Hear, Enoch, and take in these my words, for not to My angels have I told my secret, and I have not told them their rise, nor my endless realm, nor have they understood my creating, which I tell you to-day.

[4]For before all things were visible, I alone used to go about in the invisible things, like the sun from east to west, and from west to east.

[5]But even the sun has peace in itself, while I found no peace, because I was creating all things, and I conceived the thought of placing foundations, and of creating visible creation.

Chapter 25

[1]I commanded in the very lowest parts, that visible things should come down from invisible, and Adoil [7] came down very great, and I beheld him, and lo! He had a belly of great light.

(7) **Adoil.** Or, "Light of creation."

2And I said to him: Become undone, Adoil, and let the visible <u>come</u> out of you.

3And he came undone, and a great light came out. And I was in the midst of the great light, and as there is born light from light, there came forth a great age, and showed all creation, which I had thought to create.

4And I saw that it was good.

5And I placed for myself a throne, and took my seat on it, and said to the light: Go thence up higher and fix yourself high above the throne, and be a foundation to the highest things.

6And above the light there is nothing else, and then I bent up and looked up from my throne."

Would be interesting to have people of science look at this claim God made. Analyze it carefully and will find out that Dirac's ideas were incomplete but correct. In order for the source of light to exist before is visible has to exist in the invisible first. As strange as it sounds this idea goes to the heart of many scientific theories, like the antimatter ideas, or the parallel universes and even more modern ones. The explanation Enoch got was probably a version for first grade. Not able to understand a word of it but being able to agree that there indeed is visible and non-visible. As we know today the non-visible is rather more complex than the visible. The non-visible has indeed to pre-exist, therefore there is a source for visible.

That's what God told Enoch and that's what our modern science discovers.

The only problem with knowledge left in human hands is the arrogance that brings forth in humans. Most people of science tend to act like they created the universe and are above it all. I think that's the real danger why God had to isolate humans on a secluded planet.

If knowledge would lead our human nature to become humble and admire the magnificent existence maybe we would be able to use knowledge to benefit humanity. Yet, every great discovery is first used for destructive purpose. From atomic level to astronomical level people try to use every discovery to destroy others for whatever reason.

The grigori/watchers knew how to exploit this self-destructive feature humans have.

To this day humans can't find common ground to extremely small problems. This is true from family level to global level. In a family one or has to stay low or to feel that doesn't care. After a while the ego will find ways to come to surface and feel free. Unfortunately in many cases that will end up in divorce or even worse scenarios.

We are familiar by now with global level events. Diplomacy never sees a positive outcome. The ego of those in power does not try to benefit the majority or the well being of this planet.

The more science finds out the more those in power will use it to destroy not to improve or correct.

Even if global warming can't be triggered by the factors our science claims it does, we should try to protect our environment.

After all we should admit we don't have enough information about our planet regarding global warming. If one will care to read the Book of Enoch at some point in the Ethiopian version he explains how the moon can trigger extremely large effects on earth. From daily tides to monthly ones, from moon revolving on its own orbit to its one face position one can find out the effects the moon can exercise on this planet. For instance the 28 years cycle of the sun does trigger temps variations on Earth.

I recall taking a meteorology class with a professor who was considered by many of his colleges "mad". Indeed he wasn't what one would call the "classic professor". According to data he found in "secret" books, according to him, the Earth's temperature varies in cycles of 30 years 60 years, 90 years. When these cycles are complete they repeat again.

According to the Book of Enoch the cycle for sun is 28 years and for the moon the great cycle is 532 years. I wonder why is it that nobody cares to look into this aspect? Could it be because they might loose their grants? Could be because nobody has data to verify for such periods? This could be a serious aspect to look at before one is talking global warming or major changes due to the fact that we polluted the environment. After all in a form or another, in a concentration or another everything there is on this planet was there before humans did touch it. Nothing was carried from other planets or galaxies by humans. This seems to be a simple logic and should be used to change the focus of environmental issues. But political factors are much stronger than simple logic. You might be right, the atomic reactions are somewhat forced by humans, but all the chemicals are found on this planet.

As you can tell, if I would go in every detail there is to go, this book could become larger than the Bible. All I want is just to bring some information of different nature to those very few who might care. Our social system is "set" in its frame so nobody could really change that rigid frame with ease. Took me over ten years to change my mind and accept what I found out. The reason why I accepted what I found out was due to the fact that had enough scientific knowledge about the aspects I did find out. I begun to wonder, how come the "ancient" book says exactly what the modern science claims to discover?

The more I asked/searched the more I found out. Soon enough I became the village idiot and nobody knew what I was talking about.

Wasn't easy for as long as I cared about what others think about me.

Seems I want to cry on someone's shoulder again, but in fact all there is to this is to help one see how little truth is being thought out there. The education is far behind the modern research. So by being conservative, schools are teaching obsolete materials even at college level.

And like that's not enough in itself, in most cases what is being thought are scientific doctrines. Just as Loye said; orthodoxy/doctrine is part of science. Once embraced will take few decades to be cleared up from the system, if doesn't get to be "adapted" to the new system.

Again evolution theory comes to mind. For decades, in fact more than a century by now, evolution theory had many "face lifts" and is still hanging around like a mummy. These days is split in micro and macro evolution, which is nothing else but the theory of the theory, since there is no real understanding or proof for either.

Many are caught in this circle of trust, thinking that what we don't know today we will know tomorrow, if we continue to use the same strategy.

Same is true for every religion. Every religion tends to embrace science in a way or another so people will find some sort of comfort believing in God. Is not easy to go to church and listen to the interpretations that make no sense and have no proof and than go to work or school and listen to the opposite and still be sane. That's why most religions try to keep the people happy by embracing scientific theories that have no proof. The other option is indeed the option to separate the two, so people can have a split personality. One has to have one personality from nine to five and another one from five to nine and call that balance.

Religious leaders love to keep people in the frame of doctrines just as much as the educational system loves to keep people in the frame of the scientific doctrines.

Ardent defenders of evolution are ready to fight religion of any kind. Their scientific doctrine must be superior; therefore they find it worthwhile to fight for it. I know I repeat this idea, but seems it bothers me too much.

Again I feel the need to cry on someone's shoulder; for more than 20 years I did defend evolution theory. Everything made sense in my mind even if I could never see any missing links. Took really advanced biochemistry classes and biology to see evolution as impossible the way it was presented in school.

When I first understood that the DNA has conservative replication and is extremely precise or otherwise becomes lethal or impossible to reproduce, my mind start to ask serious questions about evolution as a possibility.

By now I could write at least a thousand pages why evolution or micro evolution is impossible to take place. If the micro evolution is impossible the macro evolution is just another lost theory.

The day our modern science could understand chapter 15 and 16 in the Slavonic version of the Book of Enoch, which I don't think is identical to the Ethiopian version, could be the day when we might have a better understanding of the warming and cooling patterns on this planet. In fact in the Ethiopian version is dedicated even more room to explaining the solar system and the weather impact on this planet. Chapters 60s and up are mostly about the weather pattern. Who is able to understand all that? Maybe will take few decades to understand what Enoch talks about in that text.

At the same time to understand from a scientific point of view, what could be these heavens Enoch is talking about would be something our science should focus on.

Did you just break in a hysteric laugh? You must be a scientist. That's what I use to do a while back too. Ideas like these would just make me laugh at how "naïve" some could be. One day when I ask myself about the unknown and I saw it was infinite I understood that could have all those heavens and more.

By now, maybe the big picture of how our social systems mislead us and try to keep us trapped might be obvious.

Nobody will have the whole knowledge or ability to change it any time soon, even if everybody sees its flaws. This could create a real traumatic reaction on humans. Everything they loved to believe in or loved to live by will have to be changed. All one can do is to learn the truth and adapt him/herself to it. As soon as one walks that path, one could not count anymore on friends or family for support. One has to be strong enough to be able to walk by him/herself.

The catch is that the required strength comes only after one already changed his or her thinking pattern.

Let us continue with the Book of Enoch in order to learn more about our own nature.

God did tell Enoch about His creation. Genesis has only few things about it but in essence is almost the same.

According to the Book of Enoch the man was created on Friday the sixth day. Here is what God told Enoch about human nature in Chapter 30;

[10][Friday]. On the sixth day I commanded my wisdom to create man from seven consistencies: one, his flesh from the earth; two, his blood from the dew; three, his eyes from the sun; four, his bones from stone; five, his intelligence from the swiftness of the angels and from cloud; six, his veins and his hair from the grass of the earth; seven, his soul from my breath and from the wind.

[11]And I gave him seven natures: to the flesh hearing, the eyes for sight, to the soul smell, the veins for touch, the blood for taste, the bones for endurance, to the intelligence sweetness [enjoyment].

[12]I conceived a cunning saying to say, I created man from invisible and from visible nature, of both are his death and life and image, he knows speech like some created thing, small in greatness and again great in smallness, and I placed him on earth, a second angel, honorable, great and glorious, and I appointed him as ruler to rule on earth and to have my wisdom, and there was none like him of earth of all my existing creatures.

[13]And I appointed him a name, from the four component parts, from east, from west, from south, from north, and I appointed for him four special stars, and I called his name Adam, and showed him the two ways, the light and the darkness, and I told him:

[14]This is good, and that bad, that I should learn whether he has love towards me, or hatred, that it be clear which in his race love me.

[15]For I have seen his nature, but he has not seen his own nature, therefore through not seeing he will sin worse, and I said After sin what is there but death?

[16]And I put sleep into him and he fell asleep. And I took from him a rib, and created him a wife, that death should come to him by his wife, and I took his last word and called her name mother, that is to say, Eva.

According to this we have seven consistencies and seven natures. For sure we don't understand them but we have to admit we have all of them. So our sixth sense is actually the seventh one? Well, just kidding since our science is not even close to look into these kinds of natures yet.

The most confusing to me seems to be the 12th verse/paragraph which actually suggest that besides all these other natures and consistencies human nature was made from visible and invisible.

The 14 paragraph sort of copies that idea and its reason. This is good, and that is bad. And man was told that was created only for God to see what man's feelings would be towards God.

Verse 15 is actually the key to human nature.

If one reads about human nature only in this abstract way that it was presented to Enoch, might have doubts. Yet, if we try to analyze it is just that, we have only two choices; to love or hate, and God wanted to know which one, humans will choose over the other.

One could always say that doesn't care at all, and that should be the third choice. That is not a choice at all is just a temporary status, maybe the waiting period for being able to make a choice.

Some do say "I love God". Yet others say "I hate God". These people made a choice.

Those who say: "there is no God" are still in the process of making a decision, since they have no proof either way.

Would be my pleasure to interpret all this text, but that could only destroy your joy of understanding this great Book of Enoch.

This book, as mentioned earlier, was in fact the "Bible" before there was a Hebrew Bible or a Christian Bible.

Did you ever wonder; how come a book like this didn't make it to the Bible, but Paul's letters did?

This book doesn't need that much interpretation in our times. Modern science actually helps us understand not interpret what the ancient text tells us.

As I woke up this morning and flip through the channels, as usual there were a lot of commercials. On one of them, two ladies were selling a skin product, (won't tell the name here since one might think I do advertise for it) and they were very excited of its effects on human skin. Shortly after, they started to claim that the patent is pending, but at the same time they show an ancient castle claiming that in fact this product is, few centuries, old knowledge. Immediately I thought of Book of Enoch. After all is well known that during the dark ages there weren't too many skin products so the few centuries could easily be few millenniums.

I recall reading in the Book of Enoch this text:

Chapter 8
"¹Moreover Azazyel taught men to make swords, knives, shields, breastplates, the fabrication of mirrors, and the workmanship of bracelets and ornaments, the use of paint, the beautifying of the eyebrows, the use of stones of every valuable and select kind, and all sorts of dyes, so that the world became altered.

²Impiety increased; fornication multiplied; and they transgressed and corrupted all their ways."

This text could easily connect the dots between the ancient secrets about skin products.

A while back while watching something about pyramids and people's life in that time frame they were talking of how exquisite the "make-up" in those days used to be. In fact if we look at their faces on those walls is obvious the extent of the make-up that was used.

The person was talking about the few samples of make-up they found at some sites, saying that the chemical composition is extremely complex. To this day our modern science can't really tell what was in them, but are of superior quality compared to our times modern make ups.

This morning I hear that "the patent is pending" for this totally "new" skin product.

Another aspect that can be connected to the watchers touch on humans life. For those who read the Bible and wondered about Moses' encounter with the pharaoh there might be an answer to that too.

If you recall Moses went to the pharaoh with that "magic" stick and it became a snake. Pharaoh immediately called his "magic" practicing priests and matched Moses' magic. Than this continues for few magic events, and at some point the magic practicing priests gave up claiming that their power is limited to that.

Many times I did wonder from where did the pharaoh had those priests? Who were they? How did they get their powers?

The Book of Enoch answered that question for me too.

Chapter 8
"*3Amazarak taught all the sorcerers, and dividers of roots:*
4Armers taught the solution of sorcery;
5Barkayal taught the observers of the stars, (9)

(9) **Observers of the stars.** Astrologers (Charles, p. 67).

6Akibeel taught signs;
7Tamiel taught astronomy;

8And Asaradel taught the motion of the moon,"

I did cite this text before but I just felt might be worthwhile repeating it.

These watchers, as I mentioned earlier were in fact considered angels before they corrupted human kind on earth. The only fear that came to my mind now is the fact that by now you start to have doubts about what you read. One very common objection I hear when I talk to people is;—that all this is myth. From Jesus to Adam, from angels to devils, all is but myth. It sure sounds like it if we think in within the narrow time frames or ideological frames. Indeed believing that on this planet few centuries before us there were giants and they did segregate in few different "races" or "varieties" sounds a bit out there. Not to mention, to think that on this planet there were at least 10 generations of humans that lived 800-900 years is also not easy to believe. Enoch was the 7th generation from Adam. He lived on earth only 365 years as a symbol of the days in an earthly year. But eventually after that he lived on other "planets", which could be what he called heavens.

Regarding races Enoch had a vision. Part of this vision is in chapter 88 Ethiopian version verse 12-13

12Then the white cow, which became a man, went out of the ship, and the three cows with him.

13One of the three cows was white, resembling that cow; one of them was red as blood; and one of them was black. And the white cow left them"

This is Enoch's vision about Noah, who was his great grand-son. Noah was a white man but his three sons had three different skin colors, white, red and black. The book almost immediately continues with Abraham who was 10 generations after Noah.

17Then the white cow (100) was born in the midst of them.

(100) Abraham.

[18]And they began to bite each other; when the white cow, which was born in the midst of them, brought forth a wild ass and a white cow at the same time, and after that many wild asses. Then the white cow, [(101)] which was born, brought forth a black wild sow and a white sheep. [(102)]

(101) Isaac.

(102) Esau and Jacob.

This strange vision of rather animals not humans could help one understand better the variety that human genome has. By simply telling that Isaac was a white cow makes a clear image of his color. Yet his two sons are not like him at all. One is black—Esau—and one is white Jacob. Even though Abraham and Isaac were "cows", Isaac's children were wild sow and white sheep. This is a very confusing vision indeed, since animal names are used to reflect the genetic make up for humans. There is a good genetic explanation to this "effect", which was actually presented by Mendel. At some point, Enoch claims that Noah was a white born child to a dark skin couple. Too bad I can't find right now that verse. Noah was an albino for a dark skin couple. Ten generations after the segregation took place right in F2 with his sons the effect of albino is revisited again with Abraham. Abraham married his half sister Sarah. So they had an albino too. Isaac (F1) the white cow of Abraham and Sarah married Rebecca. She was also related to Abraham's blood but one of her parents wasn't a "cow". If one follows the segregations of dominant characters versus recessive characters the phenotype is determined by the dominant character. Is known that albinism is recessive. Apparently the "cow" character is also recessive. So Isaac's children phenotype is far from being like him, one is black wild sow and the other is a white sheep. As we know by now, God was looking to "select" for the white sheep. Once selected took over 400 years of hybridization in Egypt to regain vitality after these 3 family inbreeds.

Thanks to computer ability to insert I can insert the text about Noah's birth: Chapter 105 in Ethiopian version.

"*1After a time, my son Mathusala took a wife for his son Lamech.*

2She became pregnant by him, and brought forth a child, the flesh of which was as white as snow, and red as a rose; the hair of whose head was white like wool, and long; and whose eyes were beautiful. When he opened them, he illuminated all the house, like the sun; the whole house abounded with light.

3And when he was taken from the hand of the midwife, Lamech his father became afraid of him; and flying away came to his own father Mathusala, and said, I have begotten a son, unlike to other children. He is not human; but, resembling the offspring of the angels of heaven, is of a different nature from ours, being altogether unlike to us.

4His eyes are bright as the rays of the sun; his countenance glorious, and he looks not as if he belonged to me, but to the angels.

5I am afraid, lest something miraculous should take place on earth in his days.

6And now, my father, let me entreat and request you to go to our progenitor Enoch, and learn from him the truth; for his residence is with the angels.

7When Mathusala heard the words of his son, he came to me at the extremities of the earth; for he had been informed that I was there: and he cried out.

8I heard his voice, and went to him saying, Behold, I am here, my son; since you have come to me.

9He answered and said, On account of a great event have I come to you; and on account of a sight difficult to be comprehended have I approached you.

10And now, my father, hear me; for to my son Lamech a child has been born, who resembles not him; and whose nature is not like the nature of

man. His colour is whiter than snow; he is redder than the rose; the hair of
his head is whiter than white wool; his eyes are like the rays of the sun; and
when he opened them he illuminated the whole house.

11 When also he was taken from the hand of the midwife,

*12 His father Lamech feared, and fled to me, believing not that the child
belonged to him, but that he resembled the angels of heaven. And behold I
am come to you, that you might point out to me the truth.*

*13 Then I, Enoch, answered and said, The Lord will effect a new thing
upon the earth. This have I explained, and seen in a vision. I have shown
you that in the generations of Jared my father, those who were from heaven
disregarded the word of the Lord. Behold they committed crimes; laid aside
their class, and intermingled with women. With them also they transgressed; married with them, and begot children. (144)*

> (144) After this verse, one Greek papyrus adds, "who are not like
> spiritual beings, but creatures of flesh" (Milik, p. 210).

*14 A great destruction therefore shall come upon all the earth; a deluge, a
great destruction, shall take place in one year.*

*15 This child which is born to your son shall survive on the earth, and his
three sons shall be saved with him. When all mankind who are on the earth
shall die, he shall be safe.*

*16 And his posterity shall beget on the earth giants, not spiritual, but car-
nal. Upon the earth shall a great punishment be inflicted, and it shall be
washed from all corruption. Now therefore inform your son Lamech, that
he who is born is his child in truth; and he shall call his name Noah, for he
shall be to you a survivor. He and his children shall be saved from the corruption which shall take place in the world; from all the sin and from all
the iniquity which shall be consummated on earth in his days. Afterwards
shall greater impiety take place than that which had been before consummated on the earth; for I am acquainted with holy mysteries, which the*

Lord himself has discovered and explained to me; and which I have read in the tablets of heaven.

[17]In them I saw it written, that the generation after generation shall transgress, until a righteous race shall arise; until transgression and crime perish from off the earth; until all goodness come upon it.

Well let's look first for mistakes. Not all of them but one, just to show how a wrong translation can mix up things. In verse 14 says that a great destruction will take place in one year. Right next verse claims that the one that was born will have three sons. In this very example one can realize how careful one should read these texts.

If what Enoch says here would be analyzed carefully based on Mendel's laws of segregation we could see a rather interesting genetic experiment going on to this day. An albino child was born to a couple who did suspect that could be the child of an angel. That was a scary thing according to Enoch's son. According to Enoch he was in fact Mathusala's child. Noah was the third generation from Enoch. Does that mean that the line of Enoch had the recessive for albino? This recessive gene was already present in Adam. He had two sons Cain and Abel. Abel the one who was killed was a white cow too. Eventually later on Seth was a white cow as well. Noah was the 9th generation from Seth. Maybe that's where the root of "holy cow" comes from.

In "Sci Fi in Bible" are analyzed few genetic aspects and how in fact is impossible for humans to naturally change their DNA or genotype in less than 200 generations. Any honest biologist will agree that genotype can't change in 200 generations too much. Yes, indeed we have to consider that we might still have left some blood from those watchers/giants. Maybe Goliath was the last one. In the land of Canaan where the Israelites were supposed to return to the Promised Land, there were still some giants inhabiting that place. When Joshua and other few Israelites returned from their surveying expedition, they cried to the others: "we are grasshoppers" compared to them.

Enoch on the other hand doesn't give any comparison. All he said was that the men were "exciding big". There are different kinds of giants in the book of Enoch. The ones that pick him up seem to be smaller than the watchers. The only strange aspect is that the watchers had the ability to "become" like humans. (Just a hint; what if the watchers or their offspring were the ones that played with the big rocks in the desert? Maybe they even knew what the pyramids are good for.) Now there were the other kinds of angels. They were trying to save humanity from the devastating influence of the watchers.

Chapter 9 Ethiopian version

"*¹ Then Michael and Gabriel, Raphael, Suryal, and Uriel, looked down from heaven, and saw the quantity of blood which was shed on earth, and all the iniquity which was done upon it, and said one to another, It is the voice of their cries;*

²The earth deprived of her children has cried even to the gate of heaven.

³And now to you, O you holy one of heaven, the souls of men complain, saying, Obtain Justice for us with (10) *the Most High. Then they said to their Lord, the King, You are Lord of lords, God of gods, King of kings. The throne of your glory is for ever and ever, and for ever and ever is your name sanctified and glorified. You are blessed and glorified."*

These kinds of angels or archangels will go to The Most High and ask for permission to help humanity. Sounds like our eternal fairytales. The bad ones destroy humanity and the good ones try to save humanity.

Why is this continuous game necessary?

At this point I wish I could really explain the way I see it, but somehow I would need to work with all the information I have about science, genetics and the ancient book. This is not an easy task for such book. Took me years to understand the way I do and I see how impossible would be for almost anyone to find what I say as valid, if doesn't know at least as much as I do. Maybe that's why people never understand each other's ideas the way are really told. Proof to that are the gospels.

All those who wrote about Jesus' life did observe or understood his words a bit different. In fact at some point is rather clear that none of them really understood or truly believed Jesus. He was talking resurrection, yet none of his disciples were around to see when that happens. All of them, after they saw what pain Jesus went through, were sure that Jesus is dead and will stay dead.

The woman, Mary, just went to the tomb because she loved him and was trying to mourn at his tomb. We all think, just as the disciples thought, in the back of our mind that there is no resurrection and is just a myth, because is not really easy to believe that a dead body can come back to life. That's why Mary was really surprise to see the Rabbi "alive", and Jesus' disciples were nowhere near the tomb. Without "the resurrection" being reality, there would be no church in Rome that talks about resurrection to this very day. This topic is again too big and too well debated to bring it to this book. All those who wrote about resurrection had a different prospective about it. Mistakes were made as well misinterpretations, but one aspect is clear in all the gospels. Jesus was killed and three days later he was again walking among the living for another 42 days or six weeks. This corresponds to the exact amount of time the doctors recommend to this day for body to heal after large wounds.

There are so many aspects of our contemporary life that have the roots in the times before Enoch and are reflected in his book. Not sure which one should I try to make the parallel for.

Maybe the mind is most important since our modern science tries to break into our ability to think. Again "The Sphinx and the Rainbow" comes to mind. In that book the author really did analyze the work of many people of science by comparison to each other's ideas. He mentioned few famous neuroscientists, among them Wilder Panfield who wrote "The Mystery of the Mind". Don't recall reading the book. But in Loye's book is a fragment from Panfield's book where he claims that in a human's brain: *"There is no place in the cerebral cortex where electric stimulation will cause a patient to believe or to decide".* Based on this observation he arrived to a very "out of the main stream" conclusion. He claims: "By *taking thought, the mind considers the future and gives short term direction to the sensory mechanism. But the mind, I surmise, can give direction only through the mind's brain-mechanism. It is all very much like programming a private computer. The program comes to an electrical computer from without. The same is true of each biological computer. Purpose comes to it from outside its own mechanism."*

Through this simple scientific observation, science just admits that the mind of the "biological computer" comes from outside. If science is correct on this one, is rather clear that Penfield just discovered God, but won't admit it out of fear or pride. He continues ; *"Because it seems to me that it will always be quite impossible to explain the mind on the basis of neuronal action within the brain, and because it seems to me that the mind develops and matures independently through an individual's life as though it were a continuing element, and because computer (which the brain is) must be programmed and operated by an agency capable of independent understanding, I am forced to choose the proposition that our being is to be explained on the basis of two fundamental elements."* Yet after this observation he goes back to a materialistic point of view.

An earlier quote from the Book of Enoch claimed that God told Enoch that He made human nature from visible and invisible.

One of the most famous scientists in the field of neurology observes these two natures to the point where he said: "I am forced to choose". He did observe that the mind "matures" independently. Clear scientific observations, but extremely controversial conclusions. Did he just adjust his conclusions to fit in his grant money or in his scientific frame? Only he knows that.

His observations are valid to this day and he was correct *"it is impossible to explain the mind on basis of neuronal action"*. That observation will be valid for as long as there is a human mind on this planet.

Sometimes when I make fun of some arrogant scientific doctrine defenders, mostly in the field of genetics I like to ask them; "where was your mind when you were a single cell"? Of course some don't believe they were a single cell at any point. We know by now that at some point each of us started as a single cell, but is difficult to think in that cell we already had a mind. Yes the mind was already there folded in those minuscule DNA patterns or in the entire cell. That is according to our modern science and also according to the Bible. I could see harmony right here, but I don't think that will find real confluence in science and ideology any time soon.

To see the logic of Penfield's observations one would have to admit exactly what he did observe. That the "purpose comes to it (mind) from outside its own mechanism". In Jesus wording would be "that God put His kingdom in human mind". We want the things we do because we were "programmed" to want them. We want the kingdom of God because is already "in us" in our system. All we need to do is to *mature at the same time with our mind.*

"15For I have seen his nature, but he has not seen his own nature, therefore through not seeing he will sin worse, and I said After sin what is there but death?" Maybe this is the nature Penfield claims is impossible to explain based on what science can see or understand about neuronal

action. According to modern science and to God's words to Enoch we are subject to our own unknown nature.

One of the most difficult to explain aspects Socrates was trying to explain, while on the death row was about human mind. He claimed we already know what we need to know otherwise we would not recognize it. He gives an example on the spot with a random servant present there, who according to his master didn't have any education. His example amazed the master.

I find these three sources very accurate in what they claim, but different in what humanity did conclude out of each.

But let me move on because playing in one puddle would be boring. These observations and many more of this kind are out there of decades by now. Nobody seems to be able to integrate them in the fundamental fabric of our social structure. Even if I would give thousands of examples like these ones, nobody will be able to request a fundamental change in our social structure.

Only a stringent desire to search for truth, to find it and to implement its use in our society will help humanity towards a fundamental change. Allowing the flexible mind to adjust to such change would be another immense social step. People resent and oppose change by their nature. That will be another aspect to deal with, maybe the most drastic one.

As we know change is not up to us, just happens. Ready or not here it comes. None of us knew when we were born why we were born. And none of us is quite sure when we die why we lived. Doesn't sound too "positive" from an arrogant point of view, but could be very positive if one accepts his/her own human condition.

Let's try a bit more modern genetics. Cloning is still a topic that could bring us together to fight each other for it. One wants cloning because is advanced science, others oppose cloning because alters their life.

Chapter 30 Slavonic version

"[16]And I put sleep into him and he fell asleep. And I took from him a rib, and created him a wife, that death should come to him by his wife, and I took his last word and called her name mother, that is to say, Eva."

Hmm, the mother of human kind is a clone?

Maybe cloning is as ancient as human kind, if is not, that means Eva is not the woman we think she was. If we oppose cloning we might as well oppose our human nature. Cloning seems to be a simple logic, according to this ancient text. Eva was made from some rib of Adam while he was under anesthesia. Today we are looking for special cells which are "forbidden" by law but we still don't know that the ribs are best for cloning not the stem cells. Maybe the day when science will discover that aspect, cloning will be allowed by each of us. Am I in favor of cloning? If God was in favor of it, who am I? Some argue about the soul of clones. Did Eva had a soul if was taken from a rib of some other individual? Does a branch of a rose bush have the same soul as the rose bush that came from? Hope you are not waiting for my answer to this question. I presented an extensive one in "Sci-Fi Bible".

I read through a book of genetics recently, just to see what's new in it, even though I know that the books are decades old compared to up to date research. That's another way to keep a generation down. Give them obsolete information and they will fight for that doctrine all their life, unless they care enough to educate themselves with the most recent discoveries. No wonder is so much stress in schools. Teaching old doctrines in a modern world is like forcing water to flow uphill. They will not absorb that information because "feels" out of date. By doing so the new generation starts to resent knowledge. Once that feeling is achieved there is no danger of them going back for more. In fact that's exactly what educational system does to the young minds. Thank God, we are programmed to know what we want to know and be able to ask for that knowledge.

There are so many parallels that could be made between modern science and the ancient books.

Another one, which in fact, triggered my "career" as an author, was the most difficult to penetrate, of all, to this day.

That is the phenomenon of photosynthesis.

Recently I had a long talk with someone who claims to be a Muslim. He is honest enough to admit that he doesn't read everything about his own religion. On the other hand I think is the most educated Muslim I talk to so far. I mean educated in the religion he claims to belong to. So by reading back and forth thru the suras I got to sura 36. The last paragraph reads: *"Does not man see that We have created him from a drop? Yet, be hold, he is a manifest disputer, and has set forth for Us a parable, and forgotten his creation, saying: "Who will bring the bones to life when they are decayed?" Say: "He will bring them to life Who produced them the first time, since He knows about every created thing, He who gave you fire from the green tree, so that, be hold, ye kindle flame from it. Is not He who created the heavens and the earth powerful enough to create their like?"*

There is a controversy regarding the singular and plural. "We have created him from a drop" versus *"Is not He who created the heavens and the earth,"* is not my desire to resolve this controversy, but "we" is definitely suggesting the angels Enoch is talking about. Not the watchers but the ones that helped him understand as much as a human could. The shocking text for me was: *"He who gave you fire from the green tree, so that, be hold, ye kindle flame from it".*

Would be interesting to ask any Muslim's opinion on this sentence and see what would be their interpretation. This sentence is very clear in the Bible, when God tells Adam and than Noah about the power of the green.

Our modern science on the other hand came to the conclusion that we are subject to the process of photosynthesis. Without the ability of the green to store energy, on this planet there would be no "kindle flame" for any living thing. Too bad this sura condenses so much the text in the Bible. I will not cite the Bible at this point since I did an extensive explanation in "Science discovers God" on this very topic. In fact by this simple idea many scientific projects lost their validity before they started. Nevertheless, they got paid millions for such projects that were a failure from the very inception. "They" are few groups of researchers that had extensive projects on this very topic.

Modern science did discover how the "kindle flame" takes place. The reaction to release energy to cellular level is just that. Every photon that was captured by photosynthesis can be released and used by any living system. After the photon is being used the molecule that the photon was part of changes to water and carbon dioxide. That one sentence explains what years of research found out. Most don't understand it to this day, because in order to see the parallel one need to know how exactly "the green" can kindle the flame. This is the reason why life on earth is so similar in its nature. All life forms have to be able to assimilate the same source of food, which is/are the sun/photons. The "storage facility" is the glucose molecule. Therefore, there was no evolution since that moment. Humans and algae alike eat the same "flame". If the way to "kindle life" on this planet did not change, nothing could've changed. Yet, this life kindle, or use of photons is the most complex process to be understood.

This takes us back to what God told Enoch about the moment of creation or Adoil. Chapter 25 Slav; *"¹I commanded in the very lowest parts, that visible things should come down from invisible, and Adoil* [7] *came down very great, and I beheld him, and lo! He had a belly of great light."*

When our modern science will be able to connect the dots between visible and invisible and understand from a different prospective why light sustains life, humanity will finally be closer to God. Some might claim that science already does that. Indeed, but is not using the logic of it in its research.

Very interesting explanation in Chapter 27 of how God made water in Slav version.

"¹And I commanded that there should be taken from light and darkness, and I said: Be thick, and it became thus, and I spread it out with the light, and it became water, and I spread it out over the darkness, below the light, and then I made firm the waters, that is to say the bottomless, and I made foundation of light around the water, and created seven circles from inside, and imaged the water like crystal wet and dry, that is to say like glass, and the circumcession of the waters and the other elements, and I showed each one of them its road, and the seven stars each one of them in its heaven, that they go thus, and I saw that it was good."

What our science claims about water at this point is a lot less than God told Enoch. But one aspect is clear; water is part of light, part of it visible and invisible. The seven natures of water still need to be discovered and that would coincide with the seven natures of the human make up. After all, our body is at least 75% water, would be nice to know our own nature better.

On this note, few days ago the first fuel cell bus was put in use in Norwich, CT. Finally the water became fire for our modern science. Who is to blame that Enoch knew about the fire in water yet we just found out that is true? *"which is both fire in water and water in fire"* claims God in chapter 19 verse 1 of Enoch's Slavonic version while talking about lightening.

To return for a moment to Eva to analyze that "initial sin" she made, which, according to the book of Enoch wasn't her sin. In fact God put in Adam the ability to tell apart good and bad. As we saw in that sentence, so there was no apple involved at the time God decided that death should come to mankind due to sin. Since humans do not know their nature to begin with, their tendency will be to sin, sin leads to death. Nevertheless the text claims that death should come to "you" man through the woman. This was already set up long before Eva ate the forbidden fruit. In the Ethiopian version in chapter 68-1 shows how God was allowing Eva to be corrupted by the "serpent" who wasn't a serpent at all but only a wicked "angel":

"6The *name of the third is Gadrel: he discovered every stroke of death to the children of men.*

7He seduced Eve; and discovered to the children of men the instruments of death, the coat of mail, the shield, and the sword for slaughter; every instrument of death to the children of men.

8From his hand were these things derived to them who dwell upon earth, from that period for ever."

Often my mind would wonder how did the snake/serpent talk to Eva? The above could serve as answer. There is another text claiming that indeed the watchers or angels have this ability; in chapter 19. This is chapter 17 Ethiopian version;

"1They raised me up into a certain place, where there was [28] *the appearance of a burning fire; and when they pleased they assumed the likeness of men."*

These angels had the ability to change their appearance. Eva maybe saw Gadrel as a serpent while he was talking to her. Sounds more like some sort of fairytale, but is mixed up in a book where most stories are not for little children. Just came to mind the Chinese dragon. The dragon was notorious for being able to change his appearance.

So the serpent is rather just another one of the watchers, or the angels God did allow among human kind. The intention was rather clear. Humans have to die, and there needs to be a mother to bring forth new generations of humans. Once that takes place the man has to follow the path that was pre-established that leads to death. Sin is usually the reason why humans have to die. Since we do not know our own nature, we tend to sin, that makes us vulnerable to death. That's exactly what Jesus message was. In human condition, not even Jesus was able to transcend it without death. Jesus message is rather clear, after death, if one believes and cares about God's plan there is resurrection. Well maybe that is not totally true. Jesus didn't have to die if it was only for himself. But He was actually showing the options human condition has. Enoch didn't have to die in order to see heaven and God or as is rather known, to see the other world.

Again humans were mislead by other watchers like Penemue; same chapter as above (68).

"*9The name of the fourth is Penemue: he discovered to the children of men bitterness and sweetness;*

10And pointed out to them every secret of their wisdom.

11He taught men to understand writing, and the use of ink and paper.

12Therefore numerous have been those who have gone astray from every period of the world, even to this day.

13For men were not born for this, thus with pen and with ink to confirm their faith;

14Since they were not created, except that, like the angels, they might remain righteous and pure.

15Nor would death, which destroys everything, have effected them;

16But by this their knowledge they perish, and by this also its power consumes them.

[17] The name of the fifth is Kasyade: he discovered to the children of men every wicked stroke of spirits and of demons:

[18] The stroke of the embryo in the womb, to diminish it; the stroke of the spirit by the bite of the serpent, and the stroke which is given in the mid-day by the offspring of the serpent, the name of which is Tabaet".

Lots of information was given to human kind, yet they were not supposed to read or write in order to avoid the corruption of their nature. Strangely enough, once the corruption took place, even God calls on Enoch to write and give his books, (366 of them according to Enoch himself) to his sons and make sure the books will be used by generations to come. So humans had very conflicting information. The watchers did educate them in many areas, which would lead them to self-destruction. God, on the other hand tries to explain as much as possible about human nature from a different aspect. The watchers never mentioned to humans their ability to become spiritual and to love God. The most profound "programming" of our human nature was overlooked by the watchers. We are programmed to know how to love, and to dream of peace on every level, from peace in our own family to peace for the entire humanity. The only problem is that we are also allowed to hate. That was the side the watchers banked on when they thought humanity so many aspects of self-destruction. God certainly knew about this nature that's why he was sure that humans will choose to sin every step of the way.

In chapter 80 God did advice Enoch to read to educate himself in order to learn more about humanity and its condition.

[1] He said, O Enoch, look on the book which heaven has gradually dropped down; [(86)] and, reading that which is written in it, understand every part of it.

(86) **The book which … dropped down.** Or, "the book of the tablets of heaven" (Knibb, p. 186).

²Then I looked on all which was written, and understood all, reading the book and everything written in it, all the works of man;

This set of gradual books from heaven could also be part of the Holy Grail. Did anybody read these books or Enoch was the only one so far? God clearly said to share all the information with future generations.

What a great advice for humanity! Understand who you are first. In every culture this became the key element. Jesus claimed the same thing. I recall a druid axiom that claimed; 'that one can never understand the universe if is not able to understand his own nature'. Is not easy to understand what, according to God, we can't understand. The conclusion remains from the times of Enoch to our times the same. This is what Enoch concluded:

"⁶At that time I said, Blessed is the man, who shall die righteous and good, against whom no catalogue of crime has been written, and with whom iniquity is not found."

This is definitely the most accurate, reverse idea of our own programming, which in fact is also based on the "golden rule". All of us dream of peace, as I said earlier. We are programmed, according to modern science by an outside "programmer" to want peace, to try to reach to God's kingdom. In order to gain peace, one needs to avoid any kind of crime. The irony is that in most cases people think they have to destroy the "evil ones". That's the very root of evil. No destruction of any kind is allowed in order to avoid crime in one's temporary life on this earth. Human history is based on this misconception. Wars after wars took place on this planet in the idea to force nations to one ideology. The main problem is that nobody really cared about ideology itself, they just wanted to be part of some big movement that helped them feel important in their mission. At this point, those who read what God told Moses about killing certain "nations" might not trust that what I just said is accurate. Indeed God did ask of Moses to kill certain nations. In fact God did destroy the Egyptians who dealt so unfairly with the Israelites. The reason, each and every time is the same: the civilization that needs to be destroyed is in fact a self-destructive civilization. More than any other civilization the Egyptians were very much affected by the teachings of the watchers. The great library in Alexandria is proof to their outstanding education. Nevertheless the education they had was based on magic practicing and idolatry. Such civilization is a self-destructive one, and has no long term existence.

One would say; God killed their first born in one of those cruel plagues. Indeed according to the Old Testament that took place, but only after pharaoh decided to kill the first born in among the Israelites. Moses escaped being killed, even though he was the first born. In his generation God gave back to Egyptians what they did allow for others. By saying "allow" I say that nobody could change the destructive order the pharaoh gave. Seems God allows us to make decisions and pays us back with our own "coin". The change the Egyptians had to make was way too much of a turn around in order to stop mistreating the slaves they had.

Throughout time humans made the same mistakes. They loved to kill and conquer others, in order to exploit. What is still nice about America is the fact that many fight for non oppressive ways for others. On Book TV few days ago Scott Riter was talking exactly about this aspect. He was heavily blaming the American public for allowing on their watch such crimes. Many would claim that for the price of security America should allow to take away the freedom. I think there is no "secure freedom". Unfortunately people have no time to read and inform themselves in order to be able to determine what would be the correct way to go. By the time they realize what's going on is way too late.

Most of the chapters in the sections from 60-100 of the Book of Enoch are about future events. Events that will be triggered by the choices humanity makes.

Ironically, comes to mind again "The sphinx and The rainbow" by Loye. In that book a bunch of smart minds tried to figure out how our mind can predict the future, prophets included. Being a book of science, God's existence is excluded. The explanation is extremely childish, but at least in the end the people of science did admit that they can't figure out the connection our mind makes in order to see the future. None of the experiments that were conducted had really plausible answers to how our mind can predict the future. Indeed to claim that somebody, few thousands years ago, came up by himself with the future of humanity is just too much to think of. At best, people want to know their own future for some reason, and when they try to be prophetic they will do it for money or power. The irony with the prophets that came for Israelites is that all of them were mistreated by the kings in their times. Many were killed. Jeremiah had to hide to escape. His writings were destroyed and he had to write them again. Elijah had to be taken to heaven to escape the cruelty of humans regarding his "predictions". For sure the prophets had no real reason to predict the future.

I don't think that any of the prophets really wanted to be prophets. In fact I recall Jonah or "the man of the fish" as is known in the Koran. He did whatever he could to run away from delivering the news to those living in Niniveh. Therefore in no shape or form are the prophets to be confused with the "seers of the future".

Sounds there are many whys. Why did God do this or that or the other? Nobody has a clear answer to all the whys. By reading carefully the Book of Enoch at least I got many unclear areas in the clear. Still have a long way to go if I care to know more. This search becomes really exhausting at times. I find it even more exhausting to look at all the events that do take place in our world now, thousands of years after this kind of knowledge was given to humanity. Everybody seems to see it to analyze it to talk about it, but very little can be done about it. People are under the influence of some doctrine; therefore tend to think in within the frame of their doctrine. The ones in power know how to exploit that status, so nobody would dare to affiliate to an out of frame ideology. This rigidity that is generated by the power of doctrine leads to hate among people.

That's why I think could be a good idea to have some kind of reliable forums for helping people to find the truth without having to go through a long tedious search by themselves. These forums should have a special TV channel like the Christians do. People of science and different religions or ideologies should present the truth about their ideology. Make observations or simply show the aspects that are lacking the truth or where the truth is intentionally omitted. This could be done in a civilized manner if those who present the ideology are not attached to it and they do know more than the average person knows about that ideology. The chance to corrupt is minimal if one doesn't have to defend a particular ideology or even the scientific doctrine.

As I watch on Book TV the debate between Hitchens and D'Souza, the other day, the tendency to corrupt with the intention to defend was painfully obvious. Both of them would come up with all kinds of assumptions. It was painful to watch these two people who are rather famous by now for their ideology and see how little they really know about the subject they are talking about. At some point D'Souza admitted that he doesn't know, but made the same affirmation in the name of his opponent too. Hitchens didn't move a muscle to say he doesn't know, but was clear he did not know.

Most books of science claim that the future will take science to new steps. Such claim sounds very logic to me. The problem I see with all those steps is the fact that majority of research is still working towards proving a doctrine. The doctrine of random accident regarding the formation of the universe and life on earth is obsolete by now. Nevertheless science is still searching for ways to prove it correct. I'd call that scientific stagnation.

Maybe you are not totally convinced that indeed our social system needs to change in such a way to allow people to think outside the frame of a specific ideology or religion. I could see that coming, because most people are part of their ideology with all their heart or with all their money. The war of ideologies is not really a war of ideas, is rather a war of power and influence. No religious leader will tell those who pay him or her that he or she is not telling them the truth. In some cases is because they themselves never cared to know the truth, in other cases is because pays more to be a false "prophet" than a truth speaking one.

A while back I did chat with someone in DC area, at least that's what I was told. I could tell that the person was trying to protect his identity. As we talk for a while we did realize that we come from different ideologies. Well I use to defend what he was defending, which was atheism. So for me was easy to recognize his opinion. As we went along in our chat at some point he claimed he is a scientist so he can't really afford to claim that there is God. At that point I did realize that he will never be sincere about his own thoughts. The reason I thought so was the fact that he kept coming back for more talk about my opinions. As we became more sincere in our sharing of ideas we often had moments of agreement.

For instance when I mentioned about Newton and his discoveries he was very reluctant about the fact that Newton may have used ancient books in order to discover what he did. While debating some aspects of his discoveries I couldn't resist asking him the question Newton was never able to answer.–What do you think gravity is? We know many laws of gravity, we can measure its force in many different circumstances, yet we don't know what gravity really is. What is it that triggers this force on earth?

After a while he was rather on the humble side claiming that is impossible to know so far. But science will discover what gravity is. Humanity is more than a century into using the gravitational force, yet science can't explain it. Once we open this page we continue on the same tone for a while, from gravity to electricity to their interferences. Than we couldn't stop only in his field of science of physics and mathematics so we continue with other fields of science. After we came to some sort of agreement I ask him if he would write at least one page for this book. He left without an answer. For months after that request, we weren't in touch. One day, due to some yahoo error I get some message from him. So just happens he was on line. He claimed he didn't send that message to add me to his list but I send him the message to be added to his list. In fact I did not send such message at all. I almost forgot all about our talks. Nevertheless he did recall our talks and he did ask me if I finished the book. I answered; that I still need to read some material but will be done soon. So I did repeat my request to him about that one page. He was gone again. At least he didn't try to deny it or try to twist the truth that he himself discovered while we were talking. I thought maybe he doesn't want to corrupt his way of thinking, in such case I would act the same way too.

Next time I got on line I found a message from him. Too bad I didn't save it. He was saying that he would like to write that page for my book but his identity has to remain secret. I left a message for him asking him to write the page, I don't need his identity, but I would love him to explain the reason why he would prefer to remain in anonymity. Sure enough in more than a month or so I didn't get an answer to that message yet.

Is it out of fear or out of pride that someone who used to talk in a rather honest fashion about aspects of life would choose not to be honest when comes to his public figure?

I'll try to reiterate what actually he admitted "under no pressure";

'No phenomenon is known or fully understood by human science. Lately science did manage to discover many laws of each observable phenomenon. Therefore, nobody in any field of science can claim now and maybe for years to come that the underlying phenomenon is understood. But we can use in our daily life the principles or the laws of such phenomenon'. This was in fact the main idea that he did agree with and even came up with many examples at his own will. I didn't know in this regard as much as he knew that's why I ask him to write about a page for this book.

He had mainly the same ideas I wrote about in "Science discovers God" so when I did hear even more elaborate ideas from someone who claim to be a famous scientist I was trying to get them in this book. Unfortunately for me, usually few examples are sufficient for me to see some sort of pattern. He knew a lot more than I do, but he saw no pattern at all or if he did, wasn't the pattern his ideology would favor.

While reading the Book of Enoch, some time after I was asking for these examples, one verse made me react in a very "eureka" fashion. To find the same idea I tried to explain for years now, in this ancient book was more "eureka like" for Enoch too. He claims it in chapter 92 verse 20-21 in the Ethiopian version.

"*20Who is there capable of thinking his thoughts? Who capable of contemplating all the workmanship of heaven? Who of comprehending the deeds of heaven?*"

"²¹He may behold its animation, but not its spirit. He may be capable of conversing respecting it, but not of ascending to it. He may see all the boundaries of these things, and meditate upon them; but he can make nothing like them."

Enoch is talking about humans again and their limited understanding. No matter what we will know, we will not understand "its spirit", or the real nature of the phenomenon we use. We can tap into it but we can't make anything like it. We can make a radio and tap into radio wave fields but we can't make the radio wave itself. We can tap into many other fields, use their existence in ways that are in agreement with their nature but we will never be able to make such fields. Of course, this gets a lot more complex than it seems at first glance, but what Enoch said is still true to this day.

If science is limited from this prospective, when will humanity really be able to "create" gravity without having its model first? When will science be able to produce the phenomenon of electrum without having the model of electricity? The proportions weren't mentioned only the phenomenon itself. To create gravity might be possible but to take it to universal proportions as it is now is rather out of our potent scientific abilities.

God did explain to Enoch what the electrum was and that very interesting explanation from God is in chapter 19 of the Slavonic version;

*"¹And for all the heavenly troops I imaged the image and essence of fire, and my eye looked at the very hard, firm rock, and from the gleam of my eye the lightning received its wonderful nature, **which is both fire in water and water in fire,** and one does not put out the other, nor does the one dry up the other, therefore the lightning is brighter than the sun, softer than water and firmer than hard rock."*

Is rather amazing to hear that God finds the lightning wonderful because of its nature, which is "water in fire and fire in water". So according to this explanation water has fire in its nature.

Could that be correct according to our science? With no doubt water has the most energy filled molecule, so much so, that our modern science uses its reaction to launch space shuttle. The fuel cell is based on the fact that the water molecule has an extremely potent reaction. Its reaction could develop more energy than the gas we use. Water is a different kind of fire, by the combination of its atoms the reaction is extremely exothermic or energy releaser. Its effects are even more potent than the fire itself. Again, this is an 1800 A.D. discovery, but God claims it long before it was discovered. This discovery wasn't used by science for decades. After a while, Russia tested its use, and they got the necessary fuel to leave the atmosphere with weights much greater than anybody could imagine before. So in some way God was correct to explain to Enoch about this wonderful lightning effect, water has fire in it. I wonder how the last combination really works; "softer than water and firmer than hard rock". Maybe like a laser beam? Wish I knew what science says about that, but I don't.

Another observation that seems to be out of its time is found in chapter 53 in the Ethiopian version;

"*3 And there my eyes beheld the instruments which they were making, fetters of iron without weight. (47)*"

Reading this affirmation made me think of some kind of flying saucer at least, if not some kind of jet. Seems the instruments could be anything like our tools today only better. In order to make "fetters of iron" without weight on earth could easily suggest some sort of flying system made of iron. If I never saw a plane, but I knew what iron is, by seeing some flying iron I could easily imagine that, that iron has no weight. "The chariots of God" came to mind. I think von Daniken wrote about ancient "planes" and pilots like creatures found in different murals of ancient civilizations. Maybe you might see a different idea in that quote.

This morning I got some good news from my chat friend. He is working on few paragraphs, since he said he can't condense it in one. We might get a whole page from him. In mean time I think I've already gave more than enough examples that do relate science to ancient sources of information. Finally my question about: who was the first human who wrote and read was answered by the Book of Enoch. That human was no human at all, he, Penueme, thought humanity how to read and write right in the first generations.

As funny as it sounds it does make sense. To this day every human needs to be educated in the "art" of reading and writing. The interesting part about this is the fact that a human can do such thing. Has the "inborn" ability to read and write. Maybe that's the aspect of our human nature we would rater address; how is it possible for a human mind to read and write and get the message?

Future science might answer that one.

The problem I see is the fact that our educational system still follows the ancient models of teaching. That adds a lot of stress to what is suppose to be the most exciting human activity, which is learning and discovering new aspects of life. Slowly, over time the excitement is being "killed" by stress. With the exception of some who are able to be curious enough to neglect the stress and continue to be above the rules.

Minds like Edison's or Mendel's or Newton's went way beyond of what they were thought in school or what their times allowed them to think.

One would say that they were genial minds, but those that lived in their time think they were not all that genial. As I explained earlier they had a hard time to get through the mesh of their times modern science. So by no means would the modern science neglect a true genial mind. Or would they?

Can't step too far from sarcasm when I think of how diligent the modern science was and is in rejecting what doesn't fit its own main stream limits. In this regard Dawkins might make a difference, by creating the confusion of pre-existing design and never trying to answer it in a scientific way.

The main difficulty is the superficial approach to basic understanding. Educational system has a systematic pattern in avoiding the use of logic.

I recall when I was put in front of a GC (gas chromatograph) machine, the instructor felt I will never get his point, since I said I never saw one like that before. When he saw how much I knew, he couldn't resist not repeating his question. I laughed at him, telling him the basic principle is the same, only the cover is different. He didn't get what I said, and that made me wonder how does he work with the GC?

That's the aspect educational system fails to give to students. The basic approach, that could help them understand without help every step of the way, almost at any level.

That's why Newton or Mendeleyev or Mendel did not need mentors or previous methods to rely on in order to understand the information they read in those ancient books. The fact that they did not understand fully what they discovered was because they did not have the entire information available to them.

One had mathematics, the other had biology or chemistry.

In our times we already have them together. There still are few missing links in order to be complete. That's why our modern science is still limited in each of these fields. Understanding the underlying phenomenon might be the key element for our modern science.

As I gather, from what I have read so far, seems that humanity has no chance to discover that aspect for as long as the first tendency is self-destructive.

Giving examples of modern scientific discoveries that do relate to ancient texts could go forever. Every time I read a page or a sentence I can see the relation in time. As I tried to think again about Adoil, or the light of creation, another connection entered my mind. Even though is extensively presented in "Science discovers God", the fact I did not know at that time about, Adoil, makes it even more interesting.

At that time I didn't read the Book of Enoch with the deserved attention, nevertheless I did show how everything there IS has to originate in light. By accepting $E=mc^2$ as correct, modern science agrees with the fact that all there IS, is energy. If we pay attention to the formula as well as to the atomic structure, we end up necessarily agreeing with the existence of Adoil. These affirmations at this point are more or less verifiable by our modern science.

Sure enough, the former affirmation that was made in the earlier mentioned book about nanobts just got more substance to it in the Book of Enoch. Simply, by the fact that Enoch was taken to heaven and his "nature" was changed by being able to renounce all his earthly needs. That renunciation came to him not only due to his own ability to renounce but by being accepted by God as a righteous man.

One would say that if all the needs we have on earth, in this physical body, are being lost, what is there left of our nature that would make us to desire an eternal life in heaven?

Indeed, at first glance seems that all is lost if one finds acceptance to eternal life. No hunger, no thirst, no greed, no pride, and no nothing that we are really use to live for. More like a state of admiration and tranquility for all eternity. Such transformation is rather not desirable if we don't realize that in that case we loose all our imperfections. Only due to our imperfections we have needs for an imperfect body. If we loose all our desires would make sense to become neutral to anything that could make a spiritual body imperfect. That's why Jesus did not desire any of our earthly weaknesses and he knew that doesn't pay off. If he had children with Mary all he did was to repeat the same mistake the watchers made. How could he save at all any human if he was falling in the foot-step that God himself called evil? If Paul was right, in that case Jesus had a double nature, the Son of God and the son of man. In that case he had to fulfill both in order to compare to human nature and to show the way to spiritual nature.

Would someone, able to perform miracles, care more about sex or any earthly desire? Some say that flying or jumping off a cliff feels better than anything else, even sex. How much better could feel the ability to perform miracles?

Based on the Book of Enoch the watchers gave up to their spiritual condition just to enter the human condition, or so it appears.

The watchers renounced their spiritual nature only to get involved with humans. Is rather clear that all they wanted for humanity was to destroy it anyway they saw possible. By mixing their blood into human kind they knew they might be able to destroy all humanity as God planned it. They had the ability and knowledge to "diminish the embryo in the womb", but they didn't use that knowledge for a reason. If it was only because they wanted sex so badly with earthly women they could've get rid of embryos or practice some kind of sterility for women, but they didn't. Usually in our times if a couple or even an individual is addicted to sex they will use some kind of protection. Is difficult to believe that the watchers didn't know about these other options they had, since they knew how to affect the embryo in the womb. Indeed they allowed the "hybrids" among humans knowing that they will destroy humans, which indeed they did. God decided otherwise, according to Enoch, and a flood was necessary to get rid of the genomic imperfections they put in human genome. Or maybe God did select for albino type for a specific reason. At the same time the red and black sons of Noah had to bring their genome along.

Regarding the flood event, the Book of Enoch claims the flood took place only for a certain area, more like a village. Maybe I repeated that, but just in case you skipped reading it or maybe you thought was some kind of error. I should look it up in the text but every now and then I should leave it up to your curiosity.

At the same time God told Enoch that everything was set up the way it did take place and the end of the world is planned by the 7 th millennium. But there is an 8 th millennium that is suppose to come up when time will not exist anymore. The light will be continuous and humanity will reach its intended perfection.

When Newton did claim that according to the Bible the end will come in 2026 A.D. maybe his math was accurate. Indeed around the year 2000 A.D. humanity will be about 7000 years "old".

The Holy Grail is not fully discovered yet. The nature of the second wave of discoveries will be with no doubt more of spiritual nature and might develop at a very rapid pace. As soon as our mind escapes the framed thinking development could gain an extremely rapid pace. Scientific articles like "Live forever" that explain in a very scientific, rather futuristic fashion, the effect of nanobots will become everyday science in the next few decades.

Educational system, due to its stressful approach won't be able to keep up with discipline and safety, especially in high schools. That will lead to an acute crisis for the obsolete educational system. By the time that will take place, the church will loose its influence and power, so will become an easy task to access information about church's origin and its actions of destroying ancient information. The Bible as we know it might go through a rigorous revision at the request of scientific world.

People of science, Newton like minds, who care to link the ancient information to our modern science, will be able to access sources that were taboo to public.

Only then our world will see the shift that can't deal with now. By then such shift will become a social necessity, if the drugs companies won't take over the world. In that case the world will fall back in the same old pattern of; 'history will repeat itself once more'.

If at least our science would be honest enough to admit the truth that has been discovered so far this process wouldn't have to get worse before it gets better. But too much money is at stake in research to admit that science indeed walks in the wrong direction.

The social harmony of spiritual and material world can't reach its balance among people, for as long as their minds have wrong priorities.

A professor who teaches biology without understanding exactly how the phenomenon takes place and forces his or her students to accept what they are told without proof is worse than a corrupt priest. One can change religions at any time, but is difficult to change the main stream scientific doctrine.

When our modern science will accept the truth that its roots are in fact in parts of the Holy Grail, the true journey of knowledge has begun.

If, while in schools, our children won't be told lies about Newton's apple, but tell them that in fact what Newton discovered was already written, according to him, and he rather talks about the shoulders of the giants not about the rotten apple.

But how can one tell such things if doesn't know about them or even worse, if is not allowed to present them as "reliable sources"?

Well, let's dream for a while since we have no option to be realistic about our future at this moment in time.

There is a good chance that at some point the main stream science will accept as accurate the ancient information. At that time our civilization will start a new era indeed. If that happens and when it happens new ways of thinking will have to be "implemented". Most people who belong to any ideology will resent that and try to somehow integrate their ideology in the main stream information. Hopefully that transition won't take too long and not a lot of ideologies will gain priority. The titans of Christianity will be on their way out by then. The titans of Islam will undermine their own religion by being greedy to gain power over the weak Christianity. If political power won't affect the power of any religion this transition should be a matter of few decades. From then on the dream of peace for humanity might see a new dawn.

In this new era, science will support spiritual research and the church will help with the information that was hidden or simply intentionally misleading. I can see harmony ahead of us. Can you?

At this point humanity will be only at the beginning of the stage where history finally won't repeat itself. Everything humanity will learn will be used for social growth and its own wellbeing. If you try to cut this dream at this point you might be right. Greed might have a word to say to how this harmony will take place.

Suddenly everything will be back to, same old, same old.

When Lenin started his movement he was afraid that greed will take over the communist system. Indeed the ones in power lived like kings and the ones who did the hard work fell out of grace. This well known unbalance, where the rich gets richer and the poor poorer, is in fact the crucial change. God gave life to everyone and the necessary supplies to maintain it. The greed of some, ruined the life of others, and led them to despair.

The dream gets gloomy if the greed is not in control. If the desire to overpower and control others has no boundaries the dream is dead.

How come Jesus dared to dream?

Would any human being dream like Jesus did if he knew for a fact that his or her reward is the reward Jesus get?

Some still yell that Jesus was killed and endured so much, but forget to add that his reward was immense.

Many hurt themselves just to get to heaven. But Jesus did not hurt himself he just followed the path he came to walk on. His role was to bring glory to God and he did. By hurting oneself or by killing others who don't believe what I do I will never be able to bring glory to God. The aspect that needs to be figured out about Jesus and bothers my ear rather often is the fact that Paul claims that Jesus is from the order of Melchizedek. This priest was king at the same time. This is first case in human history when a king was also a priest. Yet, this very priest failed to do the right thing when came to blessing God. Why would someone like Jesus choose to be part of his teachings? For sure this is an immense topic if one cares about the God of Abraham. I see so much work ahead of humanity in order to get all this right. The church will have to change its approach and give up on preaching and pick up on teaching the correct information. Jesus did not preach after all, he was a teacher. Would it make sense that Christianity should be thought not preached? Would it make sense that in any religion where Jesus is at its core there should be no preaching only teaching? Yet not one of the religions that use Jesus' teachings do they love to teach. Jesus engaged in debates, yet in Christianity debates are considered heresy. Every religion invented a set of rules that in the end forbids just that, which is the teaching part.

This time around knowledge will have to bring glory to God. Science will have to admit that such creation is indeed spectacular and can't be accidental at all. The never proven random accident of life on this planet should remain a past memory that helps humanity to get rid of the oppression of religion. Since a human being can't understand its own nature, sooner or later science will have to accept the design or creation that life is part of.

How arrogant can be to claim that one can possibly understand God since we don't understand our own nature?

Once these parameters of knowledge are understood and accepted, one has to accept the limits of our mind. As most people of science claim, our mind is somehow not "ours". Our neurons are not the mind itself. Their connections and stimulations are not producing the mind itself. Yet, after they conclude this aspect about the human mind, their next idea is still only about the visible and empirical methods. That's where the limitation of our knowledge might start. Not even the watchers were able to know the "spirit" of the phenomenon they were teaching to humans. That's why Newton and all those who used parts/books of the giants' grail of knowledge can't answer the critical questions. All they could help with was only the principles of any given phenomenon but not the phenomenon itself. Only the aspects that became visible in some shape or form but not the "non visible" nature. In fact they could never create what already was in existence. That goes for humans too as well as for the watchers that corrupted humanity. Well you might think this is fiction but soon enough this truth could come to light as reality. This could answer many questions humanity has about our mysterious planet and about our own condition on it.

Once we see the limitations of our scientific methods and accept them, science would have the chance to change strategies if indeed cares to reach a higher level.

Too much of my own opinion of how this could be done might sound invasive. The methods are simple. One only needs the ability to step aside from the main stream, just as Newton and most of the pioneers in science did. They will not get immediate rewards for doing so and that's the part nobody wants to deal with. Newton was a rather lucky case in the end. Mendel, on the other hand, wasn't as lucky as Newton.

How many in our modern times would be able to discover what Newton did, even if they had the ancient books?

Most people in research are there for some money or some fame.

This is harsh critique of our society indeed and its approach to knowledge, but everyday events keep me in touch with its ignorant approach.

Let's imagine for another page or so the beautiful dream most of us want to dream of, which is, life itself.

Nowhere in the Book of Enoch did I find any clear explanation of what life really is. When God explained to him the seven natures and the visible and invisible realm did not point out what is it that those seven natures need in order to be *alive*. Each of them can exist only in the unity they were suppose to be part of. The blood decomposes if is outside the living body. Well, it can be preserved for a while in our times, but is not delivering its function in that time of preservation.

The bones need to move the body and all the other natures we have. I am very much tempted to go back to the Bible for an example in Ezekiel 53, where he talks about the valley of the bones.

This book was supposed to be a miniature book, but I just can't resist when I think of so many other instances that could support this very idea. Ezekiel presents in this chapter how the dead bones come to life. The stages required in order for the dead to become alive. But still he doesn't define the "living soul" itself. Maybe once our science will be able to define life in a very complex way we might understand life better. Some already claim that explanation is in Genesis. But as one can tell by now Genesis has only small parts compared to the Book of Enoch regarding creation and living souls.

Well I just got an email from my scientist friend. Is very succinct, but I am thankful that he was willing to participate just in time for this book. Here is what he says:

"Of the many ways in which the history of science can be described, a continuous transition from ever expanding circles of understanding can be very illustrative. Consider for example how Kepler having expressed his laws of celestial motion could have been quite proud to announce how now he knew what the laws that described the motion of mars were but if asked as to the why were the laws the way they were he would have had to concede that he did not know. Many years later Newton would formulate his now famous laws of motion that not only explained celestial motion but also the motions of objects on earth and anywhere else. But again if asked where did the laws come from, for example why the inverse of the square (for the law gravitation) as opposed to the cube or other power, he would have not been able to answer such a question. Again years later Einstein discovered better laws and deeper descriptions and if confronted to what the reasons of his principles were he would have agreed that still there were questions we were not able to answer. This way looking at science more than illustrating that new questions are left unanswered describes a progression of increasing understanding that has provided us with a very detailed and fascinating account of how nature works."

"Signed; just a friend"

Do you find that amazing? Do you feel how indeed he is a scientist? I can't resist observing the same behavior as Penfield had. Oh well, seems it will take a while to happen for us. "Nature" works its way to truth in the end. So nature hopefully will set us free in the long run. Honest, I have no clue what he says about the inverse of the square, but I hope you do. In a recent chat he said that; "we him and I) extract different meanings from the same sentence. Not only was he correct for our case but I think he was right about every human being reaction to the same question.

Well I'll let you dream the dream I keep starting but I never finish; the dream about a perfect world where human condition is fully recovered to its perfection, which is the life with no pain and suffering. The life that finds harmony in its own existence and doesn't have to depend on imperfect ways to maintain it could be what God intended as perfect.

We are always dreaming of harmony, of peace and joy on this planet. This should be easy to achieve if we only knew the truth about our own nature, but we don't. In order to find out the truth one need to be willing to unlearn a lot of what we know or what we think is correct. That's the most difficult step. Right after we give up on what we know as being correct we sort of loose direction in life and feels like we lost the very core of our existence, especially if there is nothing to replace it with.

The Theory of evolution did that for many. All their religious values were ruined. Their lives had to accept this new idea of being some kind of monkey even if there was no acceptable proof for it. Over time, the then modern science came up with many so called auxiliary theories to support evolution. But all were just that, theories, because few decades later all the auxiliary theories fell apart as invalid. Nevertheless, by then the majority became convinced that evolution is a fact. Science never denied it in an open fashion because was too much to retract all of a sudden. Plus by retracting that the theory was correct science would've open the door to even greater oppression from religious leaders.

Thank God that did not happen in those times!

To this day Vatican functions only on ruling the others. Their power and riches are still well beyond they should be. Sooner or later they will have to teach the truth about God and continue to apologize just as they did few years back. The newly released inquisition files could put the cross of destruction on them.

The moon of Islam, that was the idol of Mecca, the Mecca that was ready to kill Mohamed will have to get out of the mixture with Jesus.

These will be major steps humanity still has to go through only to get rid of old doctrines.

Those who believe in the God of Abraham will have to revise carefully their way of thinking. They should never rely on preachers of any kind of doctrine if they care to know the truth.

By this immense shift, which seems to be right around the corner for each of these doctrines, our world could gain the freedom to search for truth without the obstructions each of these doctrines forced on people. Nobody will be proud anymore of being part of a doctrine.

Humanity will have to find harmony between visible and non-visible, between physical and spiritual. Only such harmony could take humanity towards its deepest "pre-programmed" desire of peace, self-understanding and honest love towards God.

The new "ideology" will allow one to ask and receive, to give and take even at the crossroads. Finally the request God had of humans might become reality and the path we were destined to walk on will be our best choice and main desire.

978-0-595-47503-2
0-595-47503-5

www.ingramcontent.com/pod-product-compliance
Lightning Source LLC
Chambersburg PA
CBHW020439290526
45785CB00002B/918